EDIBLE & MEDICINAL
SEAWEEDS

Edible & Medicinal
Seaweeds

A GUIDE TO
Healing & Nutritive Ocean Plants

TASHA GREENWOOD

Storey Publishing

The mission of Storey Publishing is to serve our customers by
publishing practical information that encourages
personal independence in harmony with the environment.

EDITED BY Kristen Hewitt
ART DIRECTION AND BOOK DESIGN BY Michaela Jebb
TEXT PRODUCTION BY Jennifer Jepson Smith

COVER AND INTERIOR ILLUSTRATIONS BY
© Jada Fitch
INTERIOR PHOTOGRAPHY CREDITS appear on
page 227

This publication is intended to provide educational
information for the reader on the covered subject and
is intended only as a resource. Any reader who forages
for wild seaweeds and chooses to ingest or use them for
any other purposes does so at their own risk; without
a 100 percent positive identification, no wild seaweeds
should ever be consumed or used in topical or any other
preparations. Water quality should also be evaluated
prior to harvesting. This publication is not intended to
take the place of personalized medical counseling, diag-
nosis, and treatment from a trained health professional.
Please consult a physician or other health professional
if needed.

The information in this book is true and complete
to the best of our knowledge. All recommendations
are made without guarantee on the part of the author
or Storey Publishing. The author and publisher dis-
claim any liability in connection with the use of this
information.

The publisher is not responsible for websites (or
their content) that are not owned by the publisher.

Storey books may be purchased in bulk for business,
educational, or promotional use. Special editions or
book excerpts can also be created to specification.
For details, please contact your local bookseller or the
Hachette Book Group Special Markets Department at
special.markets@hbgusa.com.

Storey Publishing
210 MASS MoCA Way
North Adams, MA 01247
storey.com

Storey Publishing is an imprint of Workman Publishing,
a division of Hachette Book Group, Inc., 1290 Avenue
of the Americas, New York, NY 10104. The Storey
Publishing name and logo are registered trademarks
of Hachette Book Group, Inc.

ISBNs: 978-1-63586-870-8 (paperback);
978-1-63586-871-5 (ebook);
978-1-66865-243-5 (audiobook)

Printed in China by Toppan Leefung Printing Ltd.
on paper from responsible sources
10 9 8 7 6 5 4 3 2 1

TLF

Library of Congress Cataloging-in-Publication Data
on file

contents

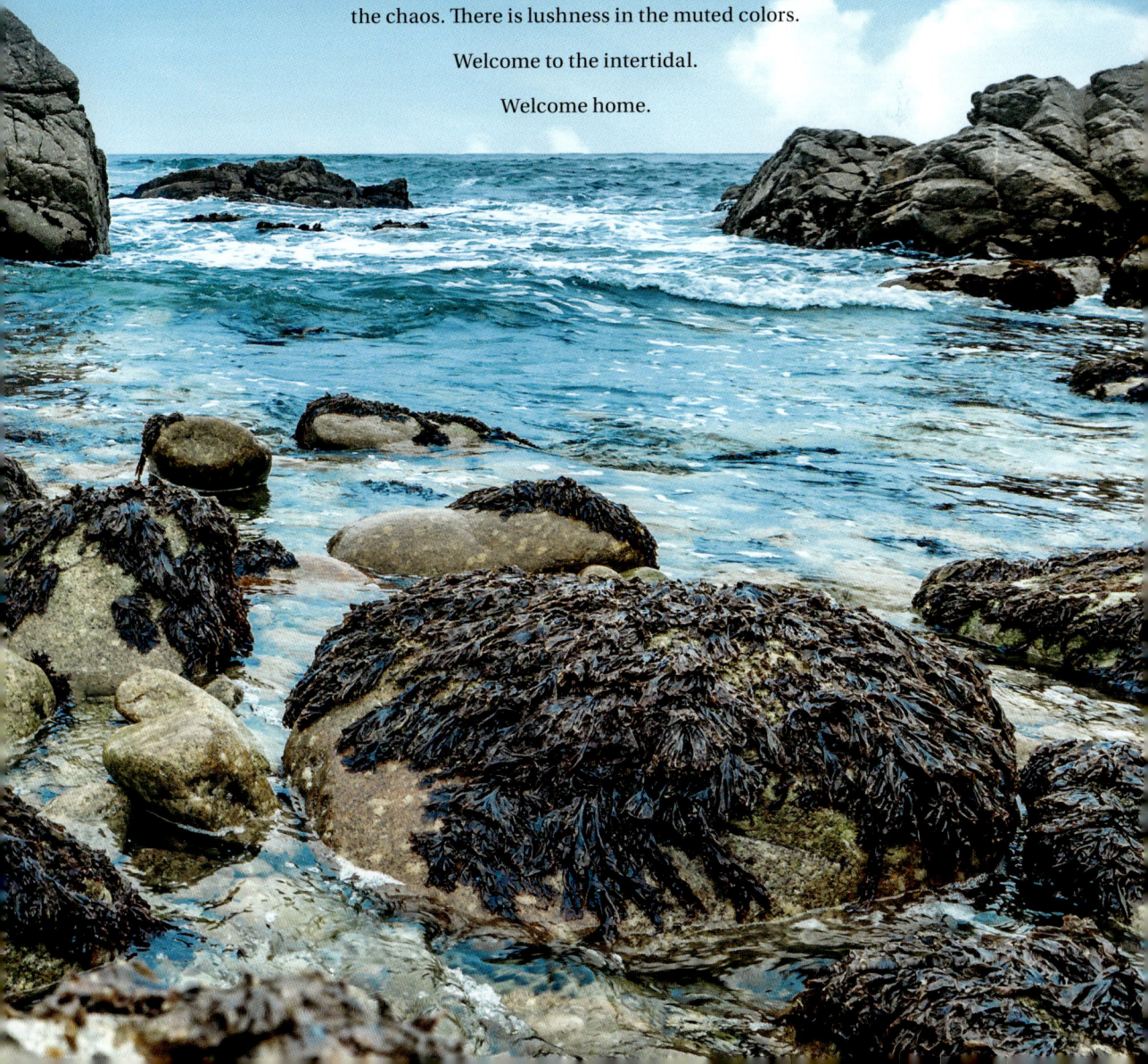

Welcome to the Intertidal

IT'S BRINY HERE. EVERYONE IS A LITTLE SALTY.
This place moves to the pulse and rhythm of the waves and tide.
The ground might be rocky, sandy, slimy, silky. There is pattern in
the chaos. There is lushness in the muted colors.

Welcome to the intertidal.

Welcome home.

Golden fronds of sargassum seaweed live free-floating in ocean waters.

Sargassum Is Home // Home Is Sargassum

The Gulf Coast of Florida is a place that was and is my home. I grew up on the same barrier island where my dad was born, and where he eventually met my mom. For more than 30 years, they rented a cozy house in a hidden oasis next to the beach amid the development of condos and giant summer homes. Mangroves and Australian pines marked the edges of my world, with a giant mahoe tree watching over everything. I was a perpetually salty, sunscreen-covered kid who insisted that my mom sew me a mermaid tail from bathing suit fabric so I could hop down the beach and fling myself into the water like some sort of gangly fish. I used to make money by entering the local sand sculpture contest and building mermaids crowned with seaweed hair and jingle shell scales. I spent hours swimming after tendrils of seaweed that harbored baby pufferfish, gently catching them in a pool of water cupped between my palms. Sometimes tiny seahorses would curl around my fingers before I released them.

The thing about living on an island (even when connected to the mainland by a bridge) and living so close to the ocean is that you always find yourself at the edges of containment and expansion. There is finite land, and a seemingly infinite watery horizon. You touch the sea, and the sea touches everything, sending salty whispers through windows, under fences, and between the pages of books. There is such a closeness with the sea, and yet it gives rise to a unique kind of isolation. I don't mean the escapism of a tropical vacation or the wealthy separatism of owning beachfront property. It's an orientation to place, to a piece of land, that you know sustains you, and knowing this land intimately is what allows you to survive even when surrounded by water. This is a place, in the subtropics and tropics, where sargassum is home.

Sargassum's Ecological Role

Sargassums (*Sargassum* spp.) are pelagic species of seaweed, meaning that they float freely in the open water rather than living anchored to a rock or the seafloor. Sargassums form large floating mats that drift through the Caribbean and Equatorial Atlantic, giving name to the Sargasso Sea. They float by use of air-filled sacs called pneumatocysts, which cluster along the golden-brown fronds that range from feathery to spiky, depending on the species and form.

When loggerhead turtle babies hatch along the coasts of Florida each year from July through October, they scramble on their tiny flippers down the sand, tumbling into the waves in search of safety. Safety for a baby turtle comes in the form of a sargassum island. The turtles' dappled brown shell blends with the seaweed coloration, and their flippers mimic the shape of sargassum fronds. They can float here securely without having to swim too hard. Food is readily available in the form of small invertebrates and fish who also hang out in the seaweed mat. Everything the young turtles need is here until they are strong enough and large enough to swim solo and protect themselves from predators. The ones who survive will eventually swim miles back to the same beach they were born on to lay their own eggs and continue the cycle.

I've lived in many different places since my childhood, and there are plenty of times when I've felt a lot like a baby turtle with tiny flippers flailing, launching myself into the surf, trusting and hoping that I'm going the right way. And seaweed has always been there for me. Seaweeds have been in the landscape, have borne witness to events large

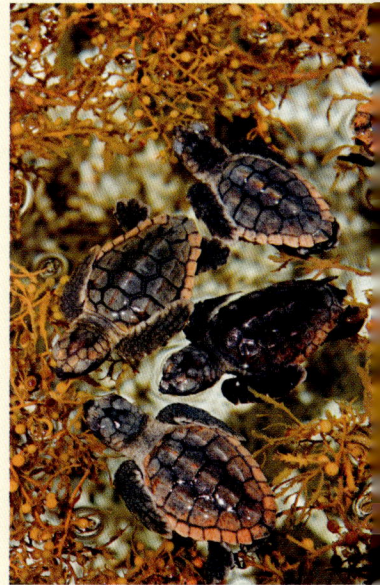

Baby loggerhead turtles finds refuge in sargassum.

Seaweed washes ashore
with the tide.

and small in my life, and have nourished my body and spirit in countless ways. I don't live on the coast right now, which is often heartbreaking and leaves me searching for water and horizons and seaweed. Amid climate chaos and a sense of instability and unpredictability permeating everything from politics to food systems, I find myself thinking about sargassum a lot. I try to remember what it feels like to be held by your island, to have the resources that you need, to float.

Sargassum has gotten a lot of negative press for washing ashore in huge quantities, covering hotel and tourist-centered beaches and dramatically shifting the coastal ecology. It's largely the excessive nutrient-rich runoff from farming and other industry that drives these huge algae blooms. While a windfall of resources for seaweed growth, this wastewater highlights the paradox we have created: We are constantly adding more and more fertilizers to depleted terrestrial ecosystems, and yet these soils can never hold the excessive chemicals that we produce and throw at them. Sometimes, it feels to me like sargassum is washing ashore and begging us to notice how much we need to nourish this land that we are living on. I think sargassum, in these disruptive displays of profusion, asks us to pause and consider questions like, "What resources do you have access to, and how are you using them?" "What is abundance?" "Who holds you?" "What is home?"

The Seaweed Economy

It's really tempting, in our desires for sustainability and sustenance, to jump from the observation of seaweed's wild abundance into blanketly declaring seaweed mariculture as the path to "food safety" in an uncertain future. Indeed, seaweed farming and seaweed products are an incredible resource economically and socially for the coasts and communities where this is a viable option. The hard truth, though, is that we have exploited many of the marine resources we have access to; communities that used to be able to rely on various fisheries find these stocks overfished and so are turning to the habitat itself (i.e., seaweeds) as a source of profit.

And the seaweed industry is booming. We'll cover many of the industrial, culinary, and medicinal uses of the specific seaweeds profiled in the following pages, so it feels important to give some context for the current moment. Seaweed has been extensively harvested and farmed off the coasts of China, Japan, and Korea for centuries and has remained an integral part of the food system through waves of cultural and political change. This has not been true, however, for much of the rest of the world. Culinary and medicinal seaweed traditions have largely been erased, predominantly via colonization and Eurocentric white supremacy. In recent decades, people have started returning to seaweed on a global scale. Restaurants of all cuisines are finding ways to incorporate sea vegetables. The pharmaceutical and wellness industries are vying to lay claim to the potential health benefits of seaweeds and seaweed extracts. In the United States, seaweed is the fastest-growing aquaculture sector, with kelp farming taking off in the cold waters of Maine, California, and Alaska. Part of me is so excited that these plants are finally getting the global attention that they deserve, and part of me worries that we will lose sight of their magic when promises of profit obscure our sight.

We know that monocropping on land undermines soil health and reduces crop quality. We also know that kelp is the singularly most reliable seaweed crop in US waters. We know that farming seaweed is undeniably more sustainable than harvesting when we consider the quantity that is in demand. I hold a lot of curiosity around the ways that we can farm seaweed alongside other mariculture (oysters and other shellfish, for example) so that we actually increase biodiversity and ocean health rather than deplete it. I wonder what happens when we think of farming and harvesting seaweed as tending our home? This book includes interviews from multiple seaweed farmers and harvesters operating at different scales, in order to share a variety of opinions and possibility models.

But I would like to invite you to meet seaweeds with more than expectations of utility, production, and profit. I would like to invite you to meet seaweeds with curiosity about what they have to teach us about adaptation and survival. I would like to invite you to meet seaweeds as you would meet members of your community on whom you depend for nourishment and care. I would like to invite you to meet seaweeds as though you are coming home.

What happens when we think of farming and harvesting seaweed as tending our home?

How to Use This Book

I wrote this book from the perspective of an herbalist and a scientist. When I left Florida, I started learning about herbs as a way to make sense of and feel held in new environments. Learning to recognize yarrow and plantain on the waysides, nibbling linden leaves in spring, and squealing with delight at the unfurling of wild rose and elderflowers has brought me into closer relationship with my body and the land on which I live. The (human) teachers who I have learned from always made it clear that plant relationships come first, and medicinal uses come second. Our lives are intertwined in a myriad of visible and invisible ways with plants. The more we prioritize and tend these ties, the more medicine plants have to offer.

Whether you are in your kitchen testing out a recipe, curious about home health remedies, diving into the chemistry of marine algae, or just appreciating the beauty of seaweed, I hope to encourage you to seek relationships with seaweeds and to know them as whole and complex beings that are far more expansive than dehydrated flecks in a grocery store aisle.

I've organized the information by seaweed type, detailing 14 of the most common edible and medicinal species. You can certainly read through from start to finish for the full seaweed immersion. Or you can hop to whichever seaweed profile you are curious about. Or you can home in on the featured seaweed farmers and companies. If you are new to herbalism, please visit the Herbal Preparations 101 appendix (page 212) for an overview of herbal medicine–making techniques, from oxymels to tinctures.

Seaweeds, like many land plants, sit in this cultural moment at a fascinating intersection of traditional use and rapid scientific discovery. Science often simply "proves" what people have intuitively and practically known for centuries. And sometimes there are new applications uncovered that are unique to the needs and challenges of this time. I've attempted to balance these sources of knowledge while also acknowledging my own positionality as a white person descended from white Europeans whose traditional uses of seaweeds I can access more easily than those of other lineages and cultural contexts. I strongly encourage you to consider where in your lineage there are coastal peoples and to learn about the seaweeds that grow on those coasts. In discovering their stories, we are brought closer to ourselves and our histories, and brought closer to home.

Eating Seaweed: Not Your Average Vegetable

One of the most common questions I get asked is, "How much of this should I eat?!" Seaweed isn't quite like spinach, kale, or other popular leafy green vegetables. The hyperconcentration of vitamins and nutrients in seaweed means that we actually can't eat large volumes of it daily. Hundreds of years ago, when diets were determined by seasonal and regional availability, before any daily multivitamins existed, humans were able to eat a lot more seaweed frequently because it was a critical source of these vitamins and minerals rather than the convenient boost it is today. With this in mind, we can get really creative about how to add seaweeds into our diets in an intentional way. We can experience the exciting flavors and unique textures of different seaweeds alongside the fruits, vegetables, and grains that are more familiar. Seaweeds are a lot like herbs in this way. You wouldn't eat an entire salad of yarrow leaves, but throwing a handful into the bowl makes for a much more exciting meal and gives a little antimicrobial and GI bitters boost along the way.

Miso soup with wakame is a delicious and healthy way to incorporate seaweed into your diet.

Seaweed and Gut Microbiomes

In some parts of the world where seaweed has always been a significant part of people's diets, we see variation in digestive enzymes that allows for greater seaweed consumption. A global study of gut microbiomes found that the gastrointestinal flora in Japanese and Korean populations contain bacteria that can produce porphyranase and agarose, the specific enzymes needed to break down the primary polysaccharides—porphyran and agar—in nori and other red seaweeds. At some point, the bacteria in these people's intestines absorbed the genes present in marine bacteria (which live on the surfaces of seaweed) and continued to replicate them, thereby allowing people to digest seaweed much more easily.

Whole Plants for Whole Bodies

There is some debate about the bioavailability of algal compounds when seaweed is eaten, and clinical research on whole seaweed consumption is limited. The medicinal properties sections cover the demonstrated and extrapolated applications and effects of various compounds found in seaweeds, with the caveat that these are largely studies done with isolated extracts. As we know, this is not how we consume the majority of our food and herbal medicine. There are enormous and often poorly understood benefits in consuming whole plants, because the different compounds often work synergistically in ways that cannot be replicated by standardized extracts. The millennia of human seaweed consumption and the revered position that these plants have held in food and medicine practices suggest that these are potent and powerful plants—and we are only just starting to understand their relationships with human health and bodies.

Best practice is to eat small amounts of seaweed frequently. The recommended dietary consumption for the average adult is approximately 3 to 4 pounds of dry seaweed per year. It doesn't sound like much, but 1 ounce of dried laver, an edible seaweed, is about the size of a frisbee; and dried seaweed expands three to five times its size when rehydrated. You can start by using it as a seasoning and sprinkling it on eggs, rice, or salads. Add it into a soup each week. Make a sauce containing seaweed flakes and freeze portions for later. Soon seaweed will become a pantry staple that fits into your cooking as easily as any other spice.

The world of identifying, eating, and supporting your health with seaweed may be very new, and you probably have lots of questions. The following pages will give you the information, recipes, and tools you need to guide your journey with seaweeds. Let's dive in!

PART 1

Welcome *to the* World *of* Seaweed

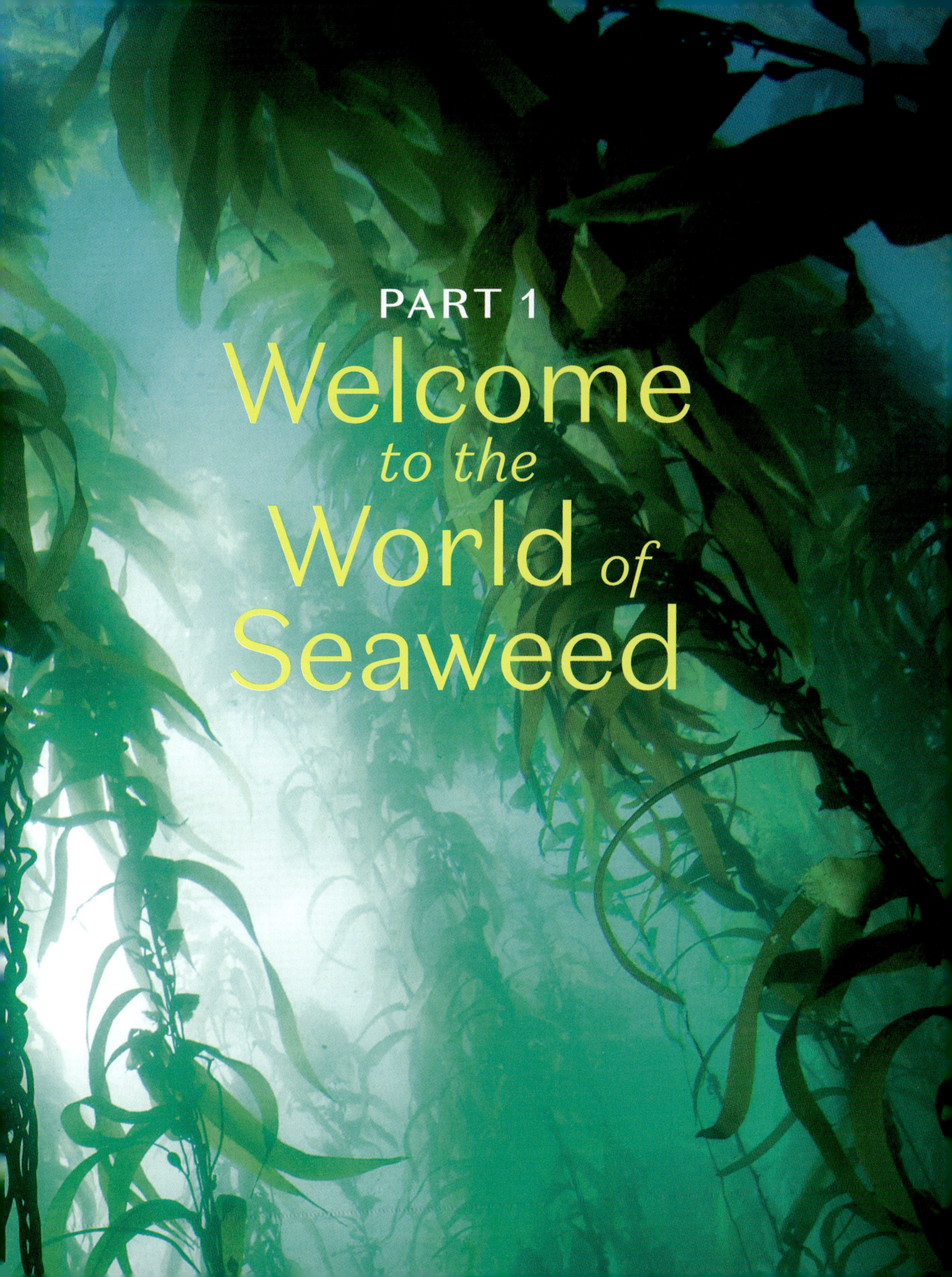

Giant kelp
(*Macrocystis pyrifera*)

CHAPTER 1

Botany Basics

The term *seaweed* is almost exclusively used to describe algae living in salt water. *Algae* is the plural of *alga*, a catch-all term referring to any simple, nonflowering, aquatic plant that contains chlorophyll but lacks vascular tissues and true stems, roots, and leaves. Seaweeds can be unicellular, but most are multicellular. *Seaweed* also specifically refers to marine algae that are macroscopic (visible to the naked eye). The microscopic spirulina is often called blue-green algae, but these organisms are actually a type of cyanobacteria, not a true algae or seaweed. However, you don't need to be a botany expert to recognize a seaweed when you see it!

Bladderwrack washed ashore might look like a miniature tree complete with branches and leaves, but it clearly belongs to the sea and has specialized features adapted to life submerged in salt water.

Seaweeds vs. Terrestrial Plants

Most terrestrial plants that we are familiar with are vascular plants. This means that they have a network of tissues called xylem and phloem that distribute water, nutrients, and carbon around the organism, much like blood vessels (vasculature) in the human body. Xylem is responsible for transporting water and water-soluble compounds from the roots to the leaves, while phloem moves sugars and proteins from the leaves to the roots. Seaweeds do not have these vascular tissues, and indeed the whole seaweed may consist of only a few different cell types rather than the 50 or so differentiated cells of typical land plants. Vascular plants take up water and nutrients through their roots (sometimes with the aid of mutually beneficial fungal hyphae), relying on soil, substrate, or the surface of another plant's roots or stem for these resources. Seaweeds instead absorb water and nutrients from the surrounding waters through all cells of their body, or the thallus. Instead of roots, most seaweeds have a structure called a holdfast, which acts solely as stabilization, anchoring the seaweed to a hard surface or deep into the seafloor substrate. Pelagic (free-floating) seaweeds, such as sargassum, do not have any holdfast structures.

Most vascular plants have stems, while seaweeds have a comparable structure called a stipe. Rather than true leaves, seaweeds have singular or multiple fronds/blades. The thallus includes the stipe and the fronds/blades together. Another specialized structure that seaweeds might have are pneumatocysts, or air-filled vesicles, that are found along the thallus and serve to provide additional buoyancy. Finally, unlike most vascular plants, which produce flowers, pollen, and seeds, seaweeds are actually closer to mosses and ferns in that they produce spores, found in structures called sori.

Epiphytic Seaweeds

Epiphytes are plants that grow on the surface of other plants and derive nutrients and water from the air rather than parasitizing a host organism. Air plants and orchids are common land plant examples. As seaweeds have evolved to take in water and nutrients from the surrounding water anyway, it is easy for some seaweed species to live epiphytic lives. Species in the *Ceramium* and *Hypnea* genera (not specifically profiled in this book) are most likely to be exclusively epiphytic, but they aren't the only ones!

MINT

KELP

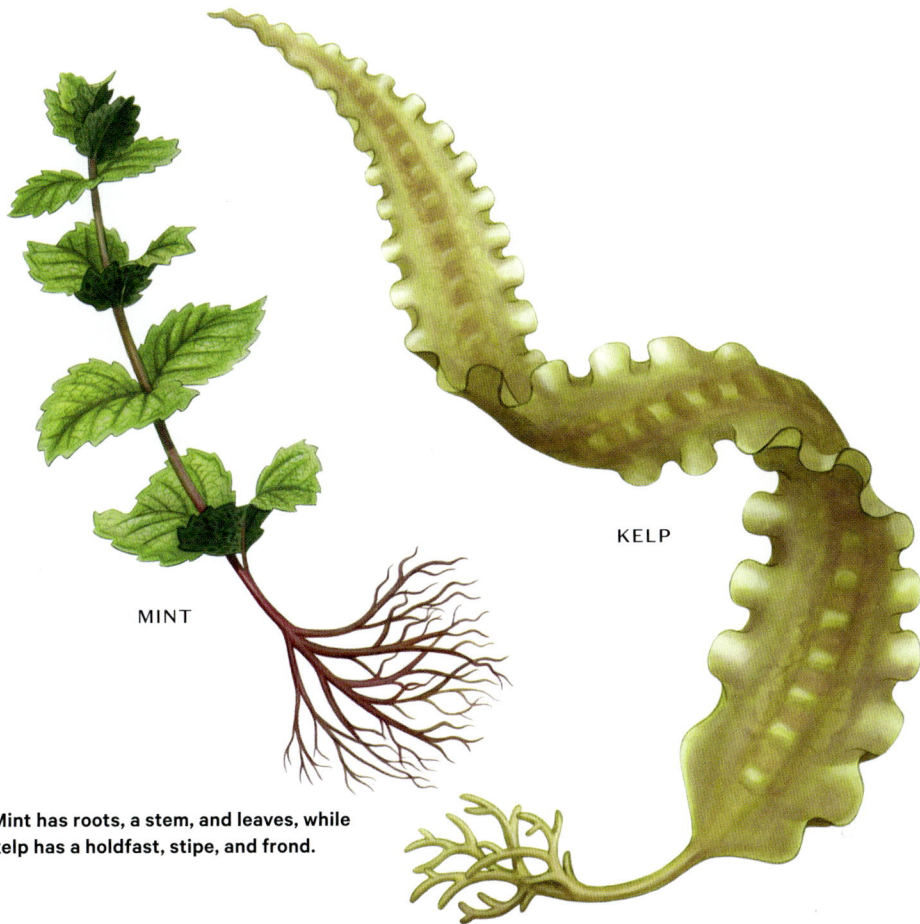

Mint has roots, a stem, and leaves, while kelp has a holdfast, stipe, and frond.

Color Groupings

There are three basic categories of seaweeds: Chlorophyta, Rhodophyta, and Phaephyta— greens, reds, and browns, respectively. Greens and reds are much older than browns in the evolutionary tree, so while they all contain chloroplasts, the functional chemistry and structure vary quite a bit between the two lineages. One of the primary differences is in the cellular structure. All seaweed cell walls are made with cellulose (a complex sugar molecule), but the cellulose is held together with different polymers (other large molecules that may include sugars, lignans, and proteins). These polymers are part of what allows seaweed to bend and twist in the waves with much more flexibility than land plants—and what allows dried seaweeds to return to their same form and texture when rehydrated. Following are further defining characteristics of each group. I'll refer to individual seaweeds throughout the book as either green, red, or brown algae.

Greens (Chlorophyta)

With few exceptions, seaweeds in the phylum Chlorophyta appear green—exactly the color that you would expect from their primary pigments, chlorophyll a and chlorophyll b. These seaweeds also contain the yellow-appearing pigment beta-carotene as well as several yellow-brown pigments called xanthophylls. While Chlorophyta is a huge phylum that encompasses a whole range of algae and freshwater species, there are only about 1,500 true marine seaweeds in this grouping. These marine species are prolific in the upper intertidal zone around the world. They are the group most closely related to land plants and use the same sugar storage molecule: starch. The polymer that holds together these seaweeds' cell walls is called ulvan. The largest single-celled organism in the world, *Caulerpa taxifolia*, belongs to Chlorophyta, with growth up to 30 centimeters long. *Codium fragile*, or "dead man's fingers," is a close second, but it grows with a branching structure. Both species have a typical multinucleate structure: The cell contains multiple nuclei and chloroplasts that move around the thallus. Green seaweeds profiled in this book are *Ulva* (sea lettuce) and *Caulerpa* (sea grapes).

GREEN SEAWEEDS

DEAD MAN'S FINGERS
Codium fragile

GUTWEED
Ulva intestinalis

BROADLEAF SEA LETTUCE
Ulva lactuca

SEA GRAPES
Caulerpa racemosa

GREEN FEATHER ALGA
Caulerpa sertularioides

Reds (Rhodophyta)

The Rhodophyta phylum contains the most distinct species, with some 7,500 named and accounted for. Red seaweeds are most prolific in warm subtidal waters worldwide and also live in temperate and cold environments. These seaweeds can thrive in deeper waters because, in addition to the standard chlorophyll, they have pigments that allow their cells to "harvest" more light as photon availability diminishes through the water column.

These pigments—phycoerythrin and phycocyanin—are fluorescent pigments specific to algae (i.e., phycobiliproteins). *Phycoerythrin* looks reddish because it reflects red light and absorbs the blues and greens, which are the shortest wavelengths in the visible spectrum and the deepest reaching in the water column. (By contrast, green algae absorb the red and orange wavelengths of light that do not penetrate very deeply in the water, and they reflect green and yellow light.)

If you are a plant nerd, you might notice that "phycocyanin" sounds very similar to "anthocyanin," which is found in land plants like purple cabbage, beets, and other purple-red fruits. Indeed, they are related pigments and have similar medicinal benefits, such as being powerful antioxidants and supporting cardiovascular health.

Either agar or carrageenan serves as the polymer that joins the cellulose in red seaweed cell walls, and the sugar storage molecule is a modified starch (similar to that in the greens). This phylum also includes the coralline seaweeds, which have calcium carbonate deposits in their cell walls, giving them the appearance of branching or crustose corals. Red seaweeds profiled in this book are *Chondrus crispus* (Irish moss), *Mazzaella californica* (rainbow leaf), *Palmaria palmata* (dulse), *Porphyra* spp. and *Pyropia* spp. (laver and nori), and *Gracilaria* spp. (Caribbean sea moss).

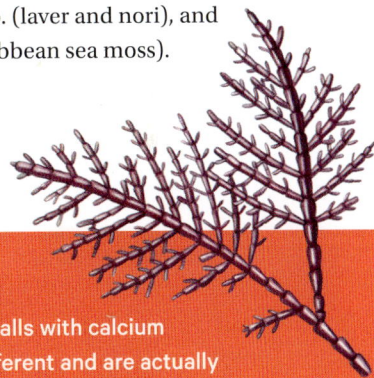

Coralline Algae

Coralline algae refers to a group of red seaweeds that build their cell walls with calcium carbonate. While true corals use the same substance, they are very different and are actually a symbiotic relationship between an animal (the coral polyp) and a plant (blue-green algae). Coralline algae can grow flat and crustlike, covering rock surfaces. Or they form wide fanlike saucers with rippled edges that stack like a colony of mushrooms. Some divide into thick and knobby branches, and others are more delicate, with articulated branches that fan out in lacy fingers. When the remains wash up on the beach, their calcium carbonate structure bleaches from pinkish red to white, looking like miniature sets of bones that clink and clack in your hands as you carry them.

RED SEAWEEDS

LAVER
Porphyra

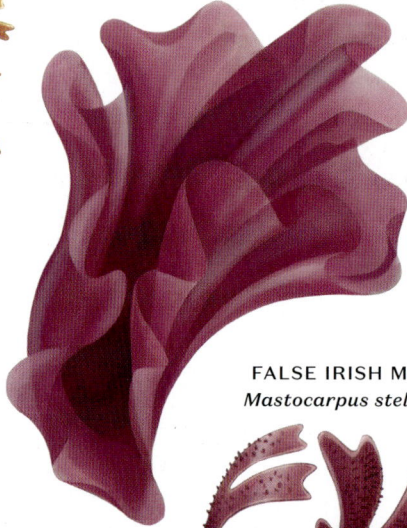

IRISH MOSS
Chondrus crispus

CARIBBEAN SEA MOSS
Gracilaria

FALSE IRISH MOSS
Mastocarpus stellatus

RAINBOW LEAF
Mazzaella californica

COMMON CORALLINE
Corallina officinalis

NORI
Pyropia

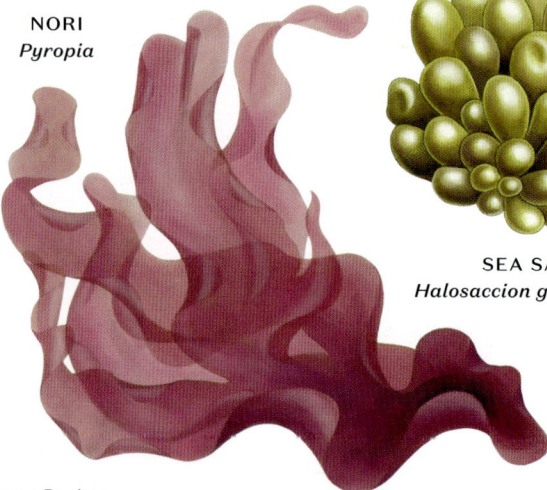

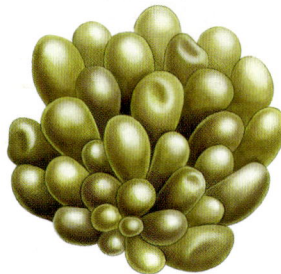

SEA SACS
Halosaccion glandiforme

DULSE
Palmaria palmata

Browns (Phaeophyceae)

Seaweeds in the class Phaeophyceae are the youngest (evolutionarily speaking) and comprise about two thousand species. These seaweeds all contain the yellow-brown–appearing xanthophyll pigment called fucoxanthin, which is coupled with chlorophyll a and chlorophyll c. Phaeophyta does not include any unicellular organisms and represents the largest seaweeds (kelps) and most prolific free-floating species (*Sargassum* spp.). Brown seaweeds are prolific in intertidal and offshore subtidal zones around the world. The polymer joining the cellulose of their cell walls is either alginate or fucoidan. Unlike reds and greens, browns store sugars via a carbohydrate called laminaran rather than in true starches. Brown seaweeds profiled in this book are *Alaria esculenta* (winged kelp), *Ascophyllum nodosum* (rockweed), *Cystoseira* spp. (bladder chain kelp), *Eisenia bicyclis* (arame), *Fucus* spp. (bladderwrack), *Saccharina latissima* (sugar kelp), *Sargassum fusiforme* (hijiki), and *Undaria pinnatifida* (wakame).

BROWN SEAWEEDS

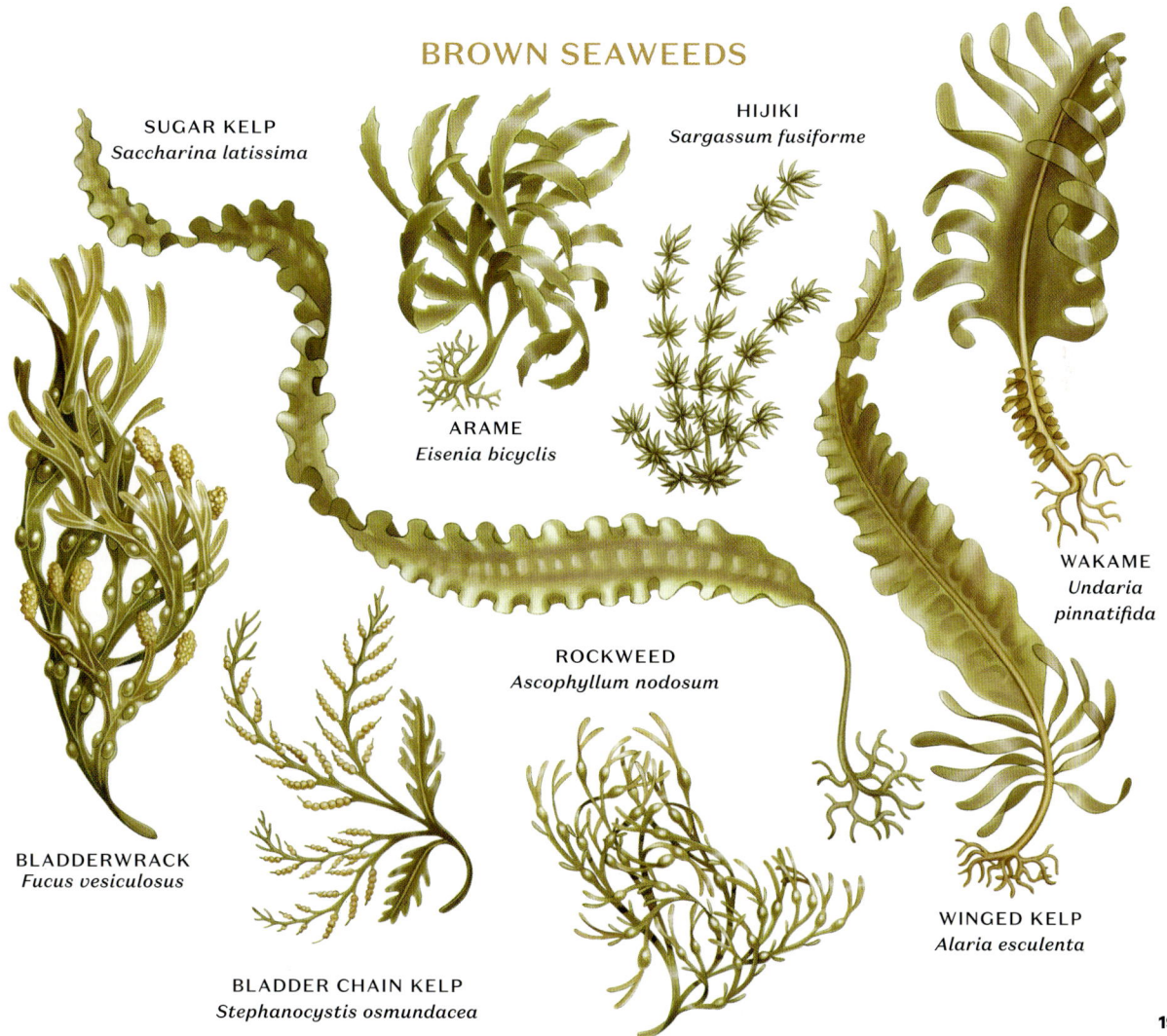

SUGAR KELP
Saccharina latissima

HIJIKI
Sargassum fusiforme

ARAME
Eisenia bicyclis

WAKAME
Undaria pinnatifida

ROCKWEED
Ascophyllum nodosum

BLADDERWRACK
Fucus vesiculosus

BLADDER CHAIN KELP
Stephanocystis osmundacea

WINGED KELP
Alaria esculenta

Reproduction and Growth Habits

For being "simple" plants, seaweeds exhibit remarkably complex life cycles and reproductive strategies. Seaweeds can reproduce sexually and asexually. Asexual reproduction in algae is via fragmentation—part of the plant breaks off, drifts to a new location, and regrows the remainder of a new clone. The generic seaweed life cycle, however, features sexual reproduction in an alteration of two generations: one haploid generation called a gametophyte, and one diploid generation called a sporophyte. *Haploid* means that each cell contains just one version of the seaweed species genome (akin to one human sperm cell), and *diploid* means each cell contains two versions of the genome (just as a human whose cells all contain genetic material from the original sperm and an egg cell). But it gets more complicated!

The gametophyte generation can produce one of two kinds of cells: Haploid spores will form structures called sporangia (which continue the haploid generation via asexual reproduction, releasing more spores to form clones of the parent organism) or gametes (which allow for sexual reproduction). Fertilized gamete cells create the sporophyte generation, which is genetically distinct from either parent. This diploid sporophyte can then produce haploid spores that are released and grow into the next generation of gametophytes. In some seaweeds, the two types of generations (haploid gametophyte or diploid sporophyte) are indistinguishable from one another. In others, the two generations look completely different, which has led to multiple instances of the same seaweed being classified as two entirely different species!

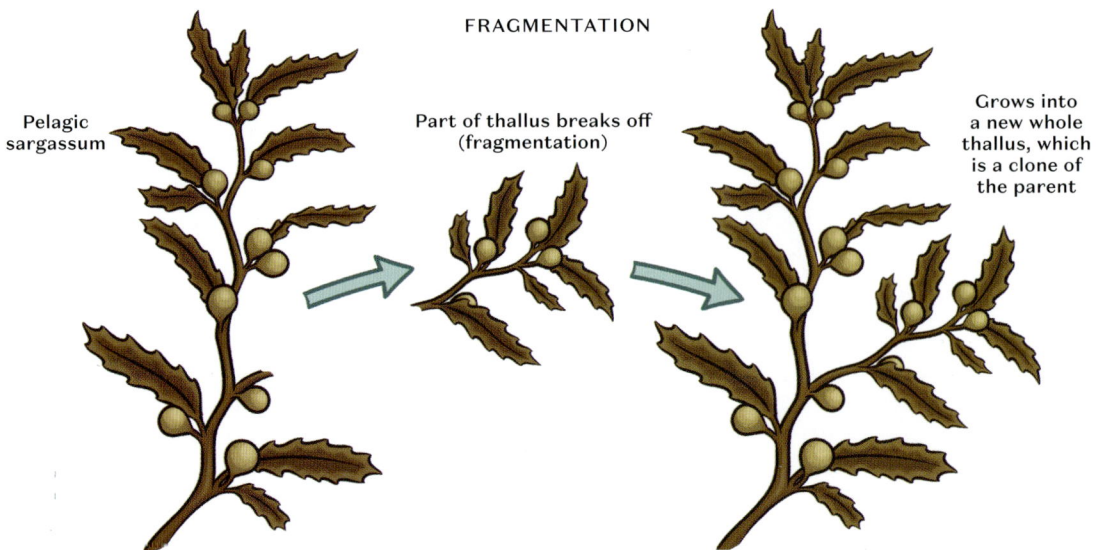

FRAGMENTATION

Pelagic sargassum

Part of thallus breaks off (fragmentation)

Grows into a new whole thallus, which is a clone of the parent

BASIC REPRODUCTION CYCLE OF ULVA

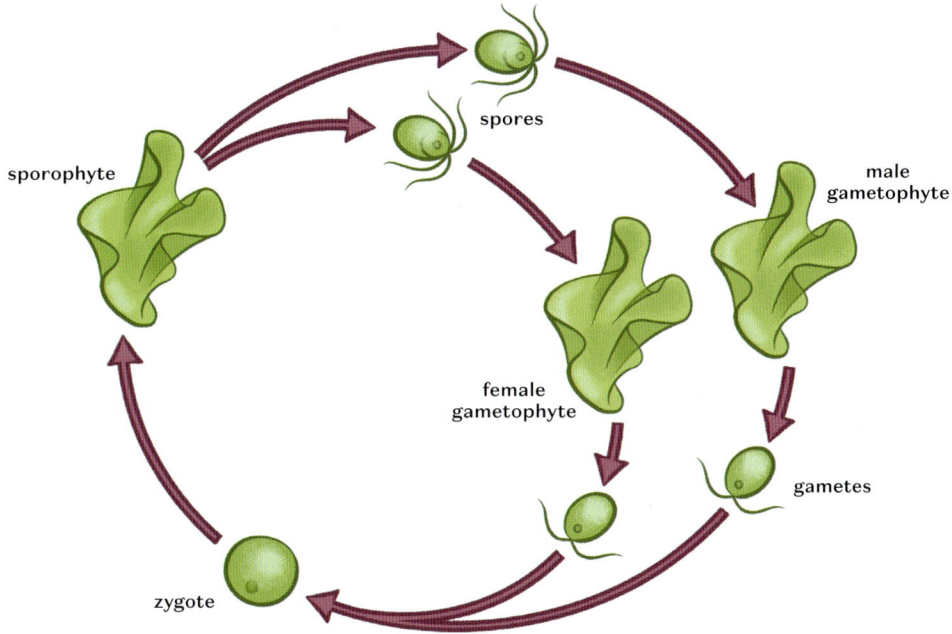

spores

sporophyte

male
gametophyte

female
gametophyte

gametes

zygote

KEY TERMS
Seaweed Reproduction

Diploid: A cell that contains two sets of chromosomes, one from each parent. All of your human body cells are diploid cells.

Gametophyte: The generation of a form of a plant that produces gametes (sperm and egg cells).

Haploid: A cell that contains just one set of chromosomes, such as a single sperm or egg cell.

Sporangia: The structure on a sporophyte that contains the spores. Groups of sporangia are called sori.

Sporophyte: The generation or form of a plant that produces spores. The frond of a fern is a sporophyte.

Greens

Green seaweeds (such as sea lettuce) reproduce asexually via fragmentation or spores from the sporangia of the sporophyte stage of the standard life cycle above. The gametes of greens have flagella, whiplike appendages used for swimming to find another gamete to fuse with, thereby forming the diploid sporophyte phase.

Reds

Red seaweeds can reproduce asexually via fragmentation or spores like greens and browns, but a defining characteristic of red seaweeds is that none of their reproductive cells have flagella, so they cannot move very fast in order to find another gamete to successfully fuse with. Some phycologists think this may be the reason for red seaweed's unusually complex alternation of generations. These complicated life cycles may also be what has allowed reds to differentiate into many more species than browns and greens. It's also the reason why farming red seaweeds is extremely difficult, and, for many species, commercially unviable.

Some red seaweeds, such as *Pyropia* spp. (nori), exhibit heteromorphic alternation of generations—meaning that the two generations look very different. The haploid gametophyte releases egg and sperm cells within reproductive structures on the gametophyte blade, which fuse to form diploid zygospores. These spores are then released and drift until they make contact with old seashells, at which point the spores attach and grow into filaments, or conchocelis, so named after the term given to these structures when they were thought to be a completely separate species. The conchocelis can then hang out indefinitely on the shell surface, reproducing asexually through spores until environmental conditions are favorable to proceed with the next generation. At that point, the conchocelis produce diploid conchospores, which are released, drift away, and settle to grow into the haploid gametophyte. Another branch of the red seaweed family inserts a third multicellular generation between the haploid gametophyte and diploid sporophyte. The generation in between is called a carposporophyte and is also diploid. Finally, red seaweeds have a third trick to address the problem of nonflagellated gametes. In 2023, the first instance of animal-mediated seaweed "pollination" was observed between the marine isopod *Idotea balthica* and the seaweed *Gracilaria gracilis*. Researchers had thought the isopod was eating the seaweed, but the isopod actually grazes on other epiphytic species and, in the process, transfers gametes between plants!

BASIC REPRODUCTION CYCLE OF NORI

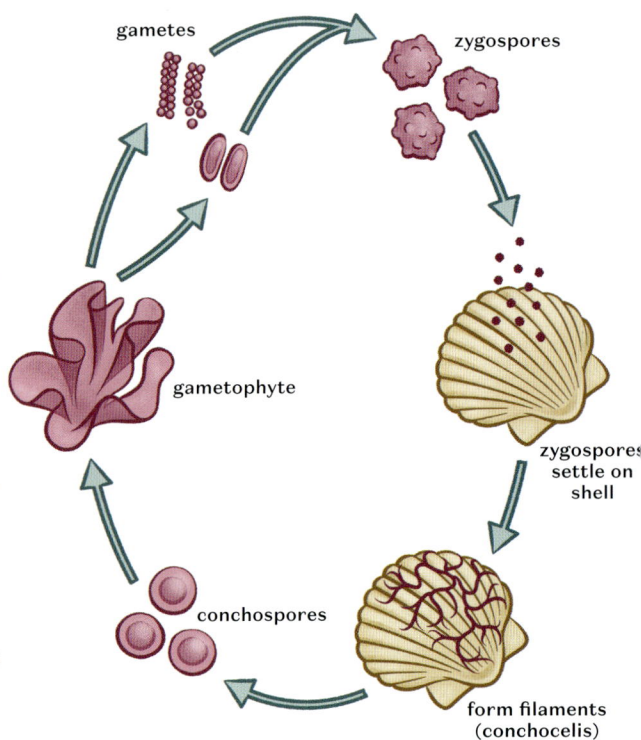

gametes

zygospores

gametophyte

zygospores settle on shell

form filaments (conchocelis)

conchospores

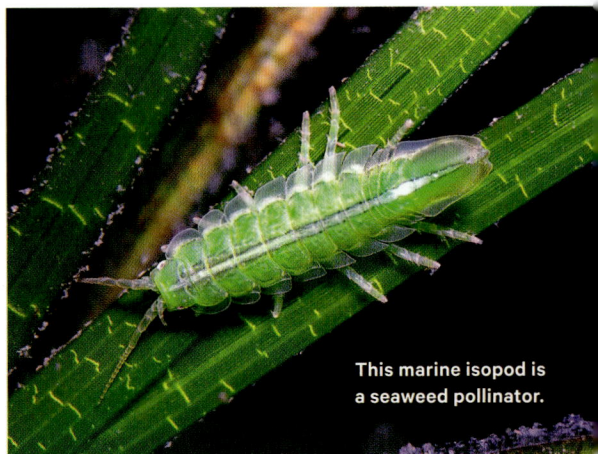

This marine isopod is a seaweed pollinator.

Browns

Brown seaweeds can reproduce asexually via fragmentation, and they have relatively straightforward life cycles. Fucoids (such as bladderwrack) have a single diploid gametophyte generation, with individual plants producing haploid egg or sperm cells, which fuse, settle, and grow into a new diploid gametophyte. Other brown seaweeds, such as kelp, exhibit heteromorphic alternation of generations (like the red seaweeds). Perhaps unexpectedly, the large fronds that we recognize as kelp are in actuality the sporophyte generation, while the gametophytes are tiny. Unlike the similar *Pyropia* life cycle described earlier, kelp is dioecious, meaning that the gametophyte phase has separate sperm- and egg-producing individuals.

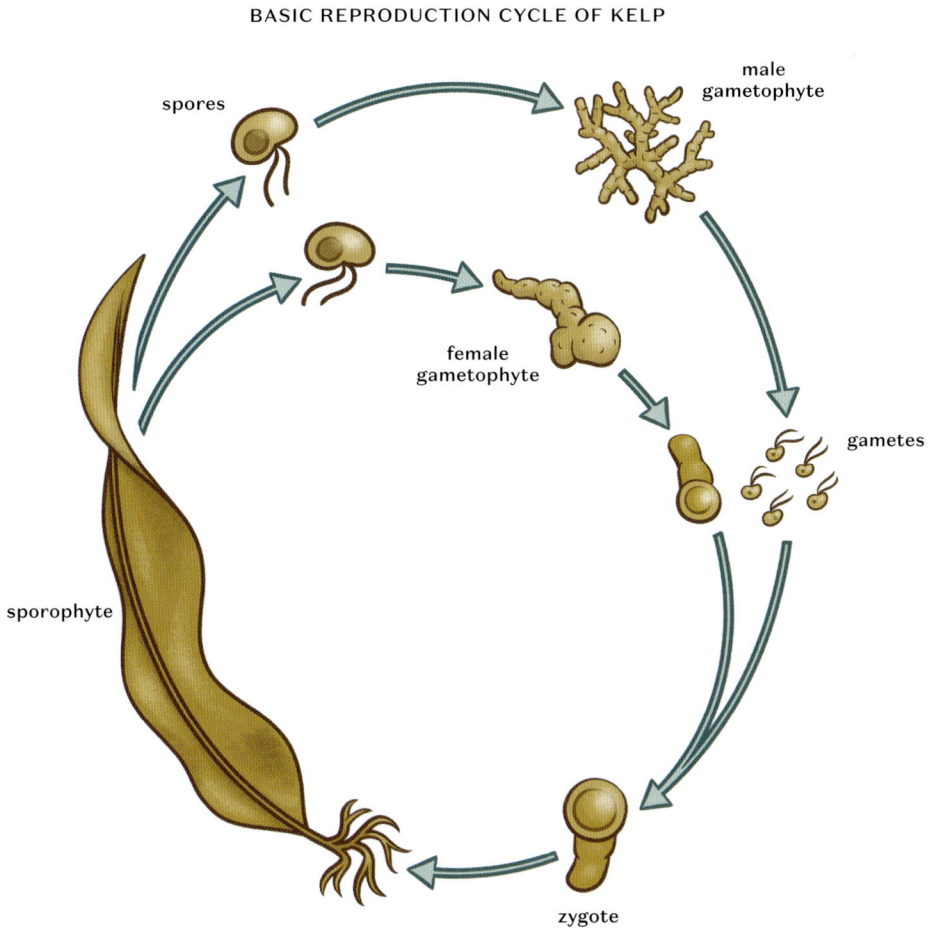

BASIC REPRODUCTION CYCLE OF KELP

spores

male gametophyte

female gametophyte

gametes

sporophyte

zygote

Wakame fronds gracefully float underwater.

CHAPTER 2
Seaweed Ecology

E cology is the field within biology that focuses on the relationships of organisms to one another and to their physical surroundings. Here I am using the term to refer to both the niches that seaweeds occupy on this planet and to the relationships we as humans have with these plants.

Rockweed drapes across the rocky intertidal zone at low tide.

Tidal Zones

Coastal locations everywhere experience two high tides and two low tides every 24 hours and 50 minutes. The tidal shift is determined by a few elements: the rotational axis of the earth, underwater bathymetry (or topography), and coastal geography. The closer a location is to the equator, the less magnitude of tidal range there is. In oceanographic terms, the part of the coastline between the high-tide mark and low-tide mark is the intertidal, or littoral, zone. Below the low-tide line and extending to the drop-off of the continental shelf is the subtidal, or sublittoral, zone. The intertidal and subtidal are the home of seaweeds.

The Intertidal Zone

As is the case with most ecosystems, the in-between places, the edges, contain the most biodiversity. The volatile and shifting environment of the intertidal means that there are many ecological niches to be filled, and that competition for resources is fierce. This leads to a profusion of different adaptations, and as each species claims its spot, we see the patterns unfold. When you go to any coastline, pay close attention. On sandy shores, you can see the wrack lines of torn seaweed, driftwood, seagrass, and other detritus marking the high-tide line and successive waves of the outgoing tide. A long sandy stretch with scalloping or with shell and rock piles reveals the prevailing currents and wave directions.

In the tidal zones of rocky shores the supertidal zone is above the high-tide line, but it may still receive some wave action. The subtidal zone is below the low-tide line and remains constantly underwater. Sandwiched in between is the dynamic intertidal zone, which is sometimes dry and sometimes submerged.

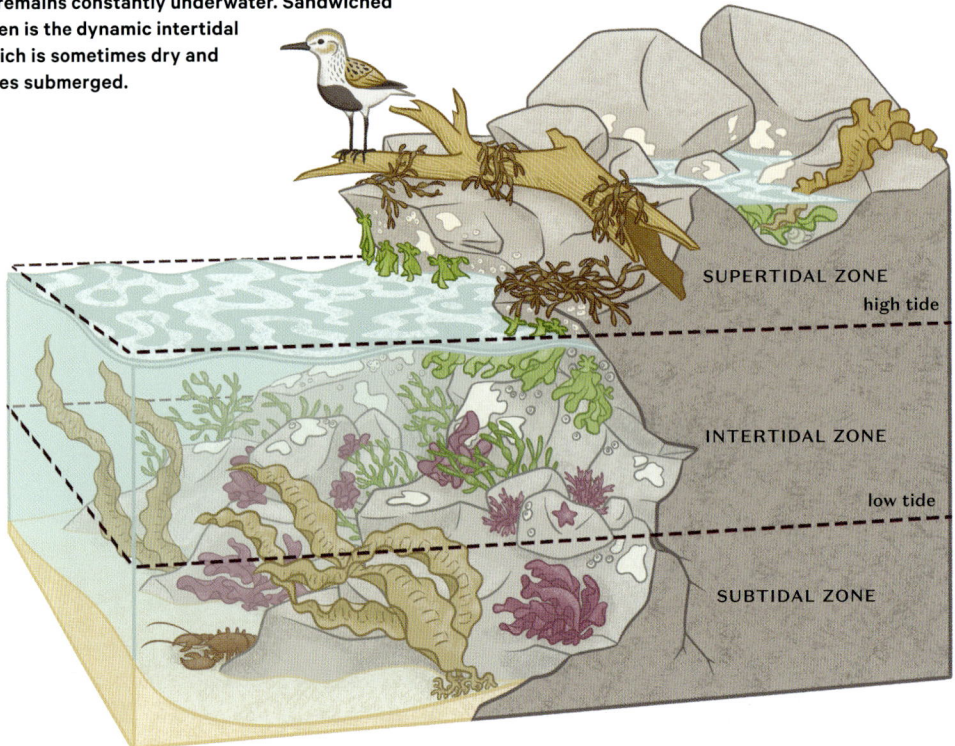

SUPERTIDAL ZONE
high tide

INTERTIDAL ZONE
low tide

SUBTIDAL ZONE

When there are rocks, you'll notice that organisms grow on them in bands. At the highest tide lines, there are often crustose algae that simply look like rock discoloration. Below that line are barnacles and mussels—often multiple species depending on your region, striated based on their competitive growth abilities and ability to resist desiccation at lower tides. And below that come the seaweeds.

In New England, where I live, brilliant green sea lettuce hangs delicately in protected spots. Then bladderwrack, rockweed, and Irish moss dominate the rocky crevasses, with other species interspersed as you move deeper into the water. The intertidal is most dramatic in higher latitudes where the tides are most extreme, and hundreds of feet of habitat exist within this liminal space.

The Subtidal Zone

The subtidal zone is characterized by constant submersion under water, even if there is significant wave action. Here we find kelps and other more air-sensitive species (often red seaweeds) filling in the spaces as we descend lower. Similar striations of individuals occur as in the intertidal, but these are based on sunlight needs rather than desiccation tolerance. Although the photic zone (depth to where sunlight penetrates the water column) extends about 600 feet, almost no seaweeds are found deeper than 90 feet. Some species, like giant kelp, have evolved large pneumatocysts to float their fronds up to the surface in order to maximize sunlight exposure while keeping their holdfasts anchored up to 100 feet below.

Species striations are visible in a rocky intertidal zone of the Pacific coast of North America. From the highest to lowest zones, we see bands of mussels and barnacles, then more delicate red seaweeds like nori, then kelps and other more hardy seaweeds among the green anemones.

Open Waters

The ocean beyond the tidal zones is referred to as the pelagic zone. As we have seen, seaweeds generally have to live in the intertidal and subtidal because they all rely on sunlight for photosynthesis in order to survive. Nutrients are rarely a limiting factor because, unlike land plants, seaweeds are not dependent on the immediate soil around them for nitrogen, phosphorus, and potassium. The movement of ocean currents is constantly refreshing the supply of nutrients in a given area. Some seaweeds, however, have evolved to simply travel with the currents, floating freely and aggregating where nutrients are most available. These are the pelagic seaweeds. Sargassum, as mentioned earlier, has solved the sunlight access problem by simply doing away with holdfasts altogether and floating on the surface of waters that would be far too deep for other holdfast-bound seaweeds to survive in. They do wash ashore across the intertidal zone but have no requirement to be confined there.

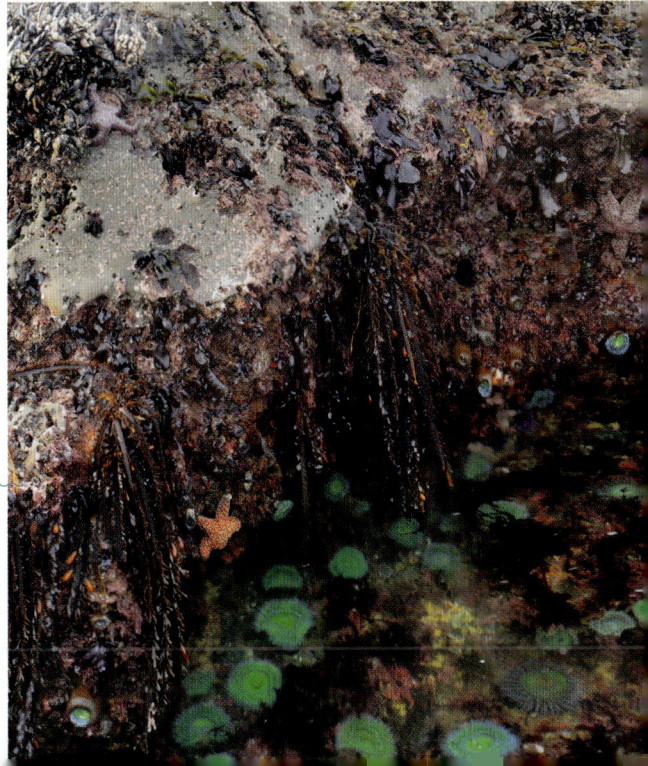

Seaweed as Habitat and Inhabitant

Seaweeds, like corals, occupy a unique position in that they are both habitat and inhabitant. The biomass of seaweeds forms incredible underwater landscapes. Kelp is considered a keystone species off the California coast because the entire kelp forest ecosystem depends on this plant for habitat structure, protection, and food. When the kelp dies back (usually due to overgrazing by sea urchins or intense sea temperature fluctuations), the entire web unravels and does not recover until environmental factors allow the kelp to return.

In the rocky intertidal ecosystem, rockweed and bladderwrack drape the rocks in shade at low tide, giving vital protection to a host of mollusks, crustaceans, and other invertebrates. It can be easy to dismiss seaweed when we only see it washed up and decomposing. Even in this condition, however, seaweed has a role to play in returning nutrients to land. For centuries, coastal people have hauled washed-up seaweed inland to fertilize and amend the soils. Decomposition and decay are just as vital to the cycle of resources within an ecosystem as birth and growth.

Seaweeds also play a key role in habitat creation on a global scale via primary production (the transformation of inorganic compounds into complex organic material), also known as photosynthesis or, to get fancy, the conversion of carbon dioxide and water into sugars (and oxygen). Although it is a bit challenging to quantify how much oxygen comes from seaweeds versus other algae, an estimated 70 percent of the world's oxygen is produced by marine photosynthesizers, and 70 to 80 percent of that is via cyanobacteria and phytoplankton. On the flip side, marine primary producers are responsible for about 50 percent of global carbon sequestration—or the capturing and storing of carbon (a naturally occurring process of the carbon cycle).

Sea otters, giant kelp, and sea urchins coexist to keep the ecosystem balanced.

A Note on Invasive Species

As with most plants and animals, the ongoing globalization of human activity combined with climate change–induced ecosystem shifts has led to an increase in "invasive" seaweed species. I am wary of using the word *invasive*—we associate this word with "bad," "damaging," and "unnatural." Per the National Invasive Species Information Center, an invasive species is one that is "non-native to the ecosystem under consideration, and whose introduction is likely to cause economic or environmental harm or harm to human health." The harm, however, is only quantified in human-centric economic terms.

Consider the perspective shift that occurs when we think of invasive species as opportunistic and resource savvy, or adaptable and resilient. As temperatures and sea levels rise, oceans acidify, and global ocean current patterns shift, marine ecosystems—and the organisms that populate them—will also change. What we see as an "invasive" species today may be the only viable species in a particular location, with the current "native" species being forced to move elsewhere. Advocating for a reduction in biodiversity caused by takeover of a single species is illogical, but I am advocating for nuance in the conversation.

Wild multiflora roses (*Rosa multiflora*) have been labeled invasive where I live in the northeastern US. People pay a lot of money to eradicate it from woodland edges and from popping up in yards and parks. But from an herbalist's perspective, this plant is an incredible gift of medicine. Roses are heart medicine, literally and figuratively. We are living through times of unprecedented global grief and rage, against a backdrop of climate apocalypse. And here is a gorgeous plant ally, the medicine that so many of our hearts need, bursting forth. Can we listen to what the wild roses are trying to tell us? And what about all the "invasive" species? What are they trying to say?

Dulse harvest

CHAPTER 3

Harvesting Techniques

& Ethical Wildcrafting

Whenever I teach about seaweed, one of the first questions I get asked is, "How do I go harvest my own?" There are so many benefits to working with seaweeds, and their beauty makes us want to experience these plants firsthand. I want you to be able to walk out to the intertidal and correctly identify and respectfully harvest your own seaweeds. Before we dive in to identifying seaweeds, it's important to talk about wildcrafting, or the practice of wild-harvesting your own plants.

Kelp fronds are harvested and laid out to dry.

Ethical Is Important

We live in a time when plants, animals, and ecosystems are being exploited, and extracted from, at higher rates than any time in human history. And I'm sure I do not need to provide you with all the examples of habitat destruction and species on the brink of extinction. Most of us are also living on land that is not ours to claim (especially here in the US). Land and coastlines were violently colonized and stolen from Indigenous peoples. The assumed right to wildcraft and white settler privilege often go hand in hand. In fact, there is a long history in the US of Black, Brown, and Indigenous people being prohibited from wildcrafting plants for food or medicine, further stripping them of resources while erasing cultural and spiritual practices. I say all of this to provide context for why wildcrafting needs to be deeply intentional in order to be both ethical and sustainable.

The Legalities

In most states, the intertidal zone is public property, and, technically, seaweed harvesting is legal. In Maine and Massachusetts, however, the intertidal zone has been privatized, and access to seaweeds here requires permission of the abutting landowners. In other states, including Texas, Florida, North Carolina, Washington, and California, property lines are determined by the mean high-tide lines, which are often disputed and can lead to confrontation over harvesting rights within the intertidal zone. Be sure to check the laws and any seasonal regulations of the places where you hope to harvest seaweed. This information can usually be found on your state government's website under "public rights" of land usage. Every state also has a department or agency for wildlife and environmental resource management, which may provide legal stipulations about seaweed harvesting.

Indigenous Inhabitants

Before you head to the coast to harvest seaweed, do a little investigating. Take time to research who historically tended these shores and seaweed beds. In many cases, Indigenous groups are still caring for these places. Some reservation lands are coastal as well. Seek out information that will tell you if there are people (individuals or groups) who also harvest from these places. This may be found through the websites of local tribal organizations or through word of mouth.

Environmental Respect

When you approach the intertidal, take some time to notice more than just the seaweeds you are hoping to harvest: What other seaweeds live here? What other invertebrates and animals are making their homes on the rocks, in the sand, or on the seaweed itself? Observe the movement of the ocean. How is the water moving in and out of the rocks, the bay, the mudflat, the sand? Do you see patterns? How does the seaweed grow in relation to the movement of the water? When you turn your attention to the seaweed you are hoping to harvest, do you notice the harvesting marks (e.g., cut fronds) of other wildcrafters? What is the population density of this seaweed? Where are they in their reproductive cycle? Are there other organisms that seem to be eating this species or using it for shelter? And finally—and this might sound silly—ask the seaweed if you may harvest them. This is the relational work of

Take time to notice what kinds of seaweeds live here besides the ones you want to harvest. What other creatures are making their homes here?

wild-harvesting, of meeting plants and getting to know them, and of reciprocity. You don't have to ask permission out loud; it is the quiet listening for the answer that matters. You will know if it's a yes. And you will hear when it's a no. If you receive a no, move on to another patch of seaweed. Or come back on a different day.

Giving Thanks

It can be really frustrating to make a whole trip to harvest seaweed and then come back empty-handed. But I encourage you to remember that capitalism and white supremacy teach us that certain humans have the right to take and to profit and to accumulate. Relational work with plants teaches us that it is not so unidirectional. If you receive a yes, offer a thank-you to the seaweed before harvesting. Gratitude comes in innumerable forms: a silent thank-you, an offering of something physical, a song, or whatever moves you in the moment. Now you are ready to harvest your seaweed.

Kelp: For seaweeds with long thalli and/or long blades, cut individual fronds or pieces of fronds from the stipe.

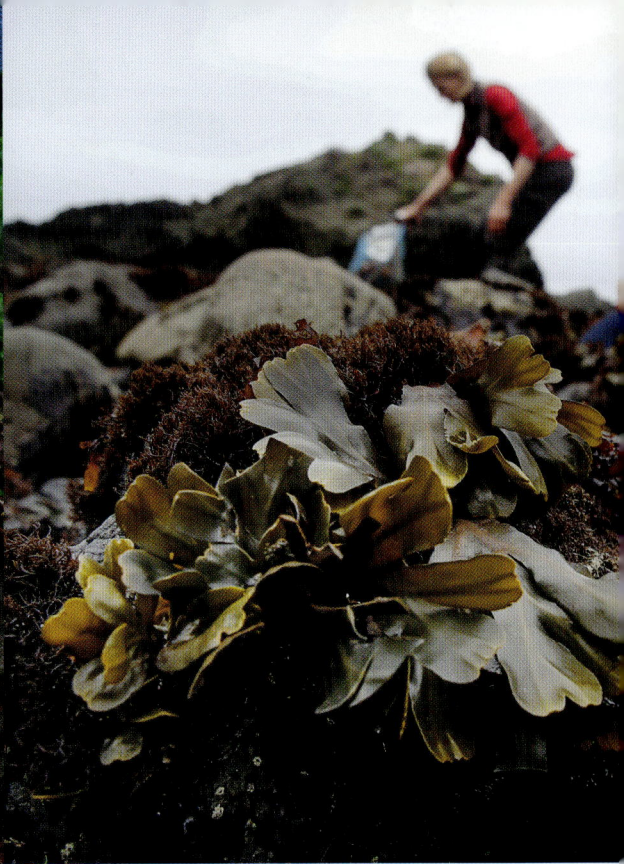

Bladderwrack (or rockweed): For seaweeds that are shorter and have a branching structure, you can trim off individual branches almost down to the short stipe growing from the holdfast.

Harvesting Techniques

Learning to identify seaweeds can be like learning to identify different types of grass—they all look similar, and they grow, more or less, in the same area. Even with a good ID book, it can be challenging to identify seaweeds because of their variety of forms and the sometimes microscopic differentiation between species. That said, technically, all seaweeds are edible. There are a few, like *Desmarestia* spp., that produce sulfuric acid, which doesn't taste particularly good and might cause stomach upset. For the most part, however, if you make an ID error and harvest a little bit of an unintended species, you won't poison yourself—it just might not taste as pleasant as you expected!

It should probably go without saying that the condition of the water where you harvest is very important. Seaweeds take up everything from the soup of dissolved nutrients, chemicals, and salts swimming around them. So pay attention to what sort of runoff or wastewater discharge might be happening nearby. As with any plant harvest, it's important to harvest only what you need and leave the rest. Everyone has different

Sea lettuce: For seaweeds that grow in flattened sheets with difficult-to-locate stipes and holdfasts, harvest a portion of the sheetlike frond, leaving part to continue growing.

Irish moss: For seaweeds that grow with tight, dense structures, take your time to locate the holdfast and tiny stipe, then harvest above that, leaving a frond or two to regrow.

ideas about the quantity (in percentages and fractions) one should harvest from a stand of plants. My method is to harvest sporadically, leaving many intact plants around a single harvested one, such that you might not even be able to tell that plants had been harvested at all.

Seaweed grows fast and prolifically but can be completely wiped out if incorrectly or carelessly harvested. Make sure to have a sharp knife or scissors and something in which to hold and carry your collected seaweed. In order for most seaweeds to keep growing, you must leave behind the holdfast, the stipe, and some frond. The photographs on this page show the growth habits of several different species.

Once harvested, you can rinse off any sand, shells, or epiphytic creatures attached to the seaweed thallus in the salt water. Never rinse freshly harvested seaweed in fresh water; the osmotic pressure difference between the seaweed cells (filled with salt water) and fresh water will cause the cells to burst and the seaweed to immediately start decomposing. Store seaweeds in a plastic bag in a cool place, or in a cooler, for transport.

Kelp dries on lines in the sun.

Drying Seaweed

Seaweed drying is easy, but it takes up a lot of space initially—before the seaweeds shrink dramatically. The easiest method is to simply hang up your seaweed outside for a day—ideally in low humidity—and let it dry until it is no longer moist or leathery and you can break it into pieces. I like using a plastic folding clothing rack, but anything nonporous works! (Avoid wood.) You can hang seaweed in the sun or shade to dry. You can also lay more delicate species (like nori and sea lettuce) out on screens to dry. This process can be a little finicky because seaweeds such as nori and sea lettuce act like sticky cling film while you're handling them—but it's worth the effort.

You might notice that the surface of your seaweed, especially kelp, looks like it's covered in a powdery mildew, but no need to worry—it's the mineral salts and sea salts drying on the surface. Once dried, seaweed can be stored in a cool place in an airtight container for years.

One of the most magical things about seaweed is that even after being stored dried for this long, it will rehydrate, uncurl, and return to its full slimy, silky glory within minutes of being submerged in water.

Harvesting from the Wrack Lines

The easiest place to see seaweed is often in the wrack lines—the swoops of detritus left behind each day as the tide recedes. But I don't recommend gathering seaweed for consumption that has washed up on the beach until you are very familiar with seaweed textures and can determine if it is freshly detached from the holdfast. Seaweeds that have been detached for several days may be starting to decompose. That said, if you are harvesting seaweed for soil or compost amendment or natural dyeing, you can definitely gather up the beached plants.

Seaweed that has washed ashore is perfect for soil/compost amendments and does not disturb the growth of seaweed beds in the water. There are no regulations about collecting seaweed that has washed ashore.

A Note on Toxicity

After a massive earthquake and tsunami in 2011 damaged the Fukushima Daiichi Nuclear Power Plant, there was serious concern about the contamination of Pacific Ocean seaweeds from potential radiation leakage. Since the initial outcry, the seaweed has been tested. It's been proven that there is no toxic radiation present in seaweeds harvested off the coast of Japan and surrounding waters. Nor was it present in seaweeds growing along the Pacific US coast, though they encounter the same waters via global ocean current circulation. The toxins we actually need to pay attention to are heavy metals—arsenic in particular. Hijiki seaweed (*Sargassum fusiforme*) has been found to concentrate particularly high levels of arsenic to the point that many government health agencies have issued advisories against consuming large amounts of it. However, hijiki is one of the primary seaweeds consumed in Japan and has long been harvested for traditional culinary and medicinal uses. Recent clinical research has also focused extensively on the health benefits of compounds found in this species. You can definitely consume hijiki safely, just in smaller amounts and less frequently than other species.

PART 2

Seaweed Herb Profiles

How to Use
This Section

The following chapters include comprehensive in-depth profiles of some of the most common edible and medicinal seaweeds. Each profile includes information for the wild-harvester, seaweed purchaser, and herbalist alike.

- **Botany & Ecology:** Each profile's botanical overview provides factual, ecological, and biological information about the seaweed at hand.

- **Medicinal Properties:** These sections provide a brief exploration of each seaweed's **energetics** (the qualities of the seaweed as it interacts with the body), **taste**, **herbal actions** (how the plant's compounds might affect us physically), and **safety** (as it pertains to use alongside pharmaceuticals, as well as during pregnancy and lactation). A deeper exploration follows with a detailed review of the clinical research and historical evidence associated with key herbal actions.

- **Medicinal Preparations:** These acute and preventive remedies include tinctures and tonics, infusions and baths, and teas and syrups. They have been tried and tested by myself and fellow herbalists—the clear instructions will allow you to do the same.

- **Culinary Recipes:** You can make these creative and nourishing seaweed-based dishes right in your own kitchen. Most recipes call for dried seaweed, so you don't have to live near the coast to acquire the key ingredient!

- **Tasting Notes:** Scientific and traditional understandings of the sweet, salty, bitter, and umami tastes present in seaweed will guide you in using seaweed in the kitchen or in medicine.

- **Seaweed Science:** These sections provide closer examinations of the chemical compounds that give seaweeds their medicinal qualities.

Guide to Shorthand in Herbal Preparations and Recipes

This shorthand is used throughout the following recipes. These elements are noted for each extract:

- State of the plant material—either fresh, wilted, or dried

- Ratio of herb weight to solvent volume (e.g., a 1:5 tincture means that there is 1 gram of herb for every 5 milliliters of alcohol)

- Metric units (these are the global standard, and conversion between grams and milliliters is easy and simple)

- Percentage of alcohol needed (e.g., 50%)

- Recommended daily dosage range in milliliters (1 standard dropperful is 1 milliliter, so measuring dosages is easy!). For some preparations that are more foodlike, dosing is given in teaspoons or tablespoons.

Some recipes feature measurements in US standard measures (e.g., tablespoons, cups) and some, specifically herbal formulas, feature measurements in "parts." I use "parts" as a measurement to give you a lot of flexibility with how much you want to make at a time. "Parts" refers to fractions of the total volume. For example, if you want to make 1 ounce of a formula that is 2 parts plantain and 1 part calendula, divide the total weight (1 ounce) by the total number of parts (3) to find the weight of each part (⅓ ounce). Now it's easy to calculate that you need ⅔ ounce plantain and ⅓ ounce calendula to make your formula!

For Understanding Seaweed's Medicinal Properties

Adaptogen: This herbal action refers to plants that help our bodies receive, metabolize, and adapt to stressors more easily.

Adjunct therapeutic: A common term, this refers to something supplementary, or auxiliary, as in clinical research or historical evidence that suggests that this seaweed is beneficial for the treatment of a particular pathology when used alongside or in conjunction with other allopathic (i.e., conventional) medications or interventions.

Dose-dependent: The principle that a medicinal plant's beneficial effects are contingent on the dosage consumed. Some plants may also have different effects at high versus low dosages.

Energetics: Herbs demonstrate qualities as they interact with our bodies. Western herbalism uses the axes of hot/cold, damp/dry, tense/lax, atrophic/stagnant, and excess/deficient to describe tissue states in the body. A cough with a copious white phlegm, for example, is a cold and damp presentation. The energetics of herbs can sometimes be obvious: The warming (even hot!) nature of ginger is readily apparent when you eat it. Others are more subtle: It can be hard to tell that linden is moistening until you chew the leaf for a few minutes to release the mucilaginous compounds it contains. Finally, some herbs' energetics change depending on the preparation: Peppermint is warming and helps break a fever when used in hot tea, but it is delightfully cooling and refreshing when served cold over ice in summer. The third context of energetics in herbalism refers to the subtle, spiritual medicine of plants. Folklore, myth, and magic come together to offer endless ways that we can interact with the energy of plants. Flower essences are the most common type of energetic herbal medicine; see Appendix 1 for more information and instructions to make your own.

Fucoidan: Fucoidans are polysaccharides found in all brown seaweeds and provide many medicinal benefits.

Herbal actions: A plant's chemical compounds have effects on our bodies. The use of terms like *antiviral*, *anti-inflammatory*, and *hypoglycemic* helps herbalists work with regulatory laws in the US, laws that prohibit herbalists (and other unlicensed healthcare professionals) from using prescriptive words like *prevents*, *cures*, or *treats*, to name a few (as in "prevents a cold," "cures arthritis," or "treats high blood sugar"). We can overlay actions onto energetics (see Energetics in this list) and describe plants as having qualities like "a warming and drying expectorant" or a "cooling and moistening anti-inflammatory."

Hydrocolloid: Hydrocolloids are polysaccharides and proteins that form a gel when in the presence of, or when mixed with, water.

Hypolipidemic: An herbal action referring to lowering cholesterol.

Immunomodulant: Immunomodulation is synonymous with immune system regulation. This herbal action refers to a quality of helping to calm a hyperactive immune response and stimulate an underactive one. This is particularly helpful in the case of inflammation, when our bodies need to mount a healthy response to destroy pathogens or heal tissue damage, but we don't want the inflammatory immune response to go unchecked and cause more damage than repair.

In silico: This clinical research term refers to studies done using a computer model or algorithm rather than living cells or whole organisms.

In vitro: This clinical term refers to studies conducted in a laboratory petri dish or test tube.

In vivo: A clinical term, this refers to studies conducted on live organisms—animals or humans. In vivo studies generally carry more weight than in vitro ones, but in vivo clinical trials on humans involving seaweeds are still relatively infrequent.

Nutritive: This herbal action refers to the nutritional value of a plant and connotes that the plant can be used in a food-as-medicine context.

Phenolic compounds: Phenolic compounds include the volatile oils that make many plants smell delightful but, for the plant, usually serve to repel herbivores, prevent infection, or heal damage. Seaweeds are not exactly aromatically perfumed, but the phenolic compounds in their tissues are the source of much antioxidant action and other medicinal benefits. Also called secondary metabolites, these molecules are formed by a plant's secondary metabolism, the processes by which plants adapt to meet disease, predation, and other stressors. (Primary metabolism, in contrast, covers the processes that allow a plant to germinate, grow, and produce seeds.)

Phycocolloid: A phycocolloid is a hydrocolloid (see facing page) found in seaweeds (the prefix *phyco* refers to seaweed). These compounds are the source of many of seaweed's medicinal benefits and are sought after in most seaweed-extract industries.

Polysaccharide: This is a large carbohydrate molecule comprised of multiple ("poly") small sugar-type molecules (saccharides) joined together. It functions as both a structural component of plants and a storage molecule for starches and sugars. You will also see the term *sulfated-polysaccharide*. All this means is that the molecule has extra sulfate groups attached to the main structure—making the primary seaweed polysaccharides (carrageenan, fucoidan, agar, and ulvan) the source of so many medicinal benefits.

Tonic: In this context, *tonic* refers to plants that can be taken daily and/or long term for general well-being and support of multiple body systems.

CHAPTER 4

Irish Moss

Chondrus crispus

I rish moss is a magical plant to me. I moved from Florida to New England for college, arriving in Boston having never seen snow. I was fascinated by the northeastern shores of the US and wondered how it was possible for organisms to survive when the shore was covered in winter ice. Twice each day the dramatic tides pulled back from the rocks, exposing the clear striping of barnacles and seaweeds perfectly following the rocky contours. There were swaths of dark brown and green seaweed, and reddish ones, too, in big patches. Some of these had a shimmer to them, a subtle but unreal iridescence that made me wonder later if I'd actually seen them. Growing lush and full, a magnified carpet of moss—Irish moss reminded me again and again what it is to survive and thrive on the face of a rock that is pounded by waves, what it is to hold such strong structure yet be soft and pliable to the touch. Like the *Mazzaella* species I'd encounter years later in the Pacific, they emit a sparkle, so to be in their presence is to be winked at from across the room by someone you love—you feel seen, your presence acknowledged.

COMMON NAMES

Irish moss, carrageen moss, carragheen, carraigín

Botany & Ecology

Irish moss (*Chondrus crispus*), a red seaweed, gets its name from its mosslike growth habit and long history of use as food and medicine along the coasts of Ireland. It grows on rocky shores just above the low-tide mark, as it doesn't love being exposed to air and direct sunlight for too many hours in a row.

Small and Tenacious

Its thallus is short, and it branches quickly into tufts of fan-shaped fronds, which are dark reddish brown with violet iridescence at the tips. It's kind of shrubby, like the way raspberry and rose brambles can form a dense hedge beneath taller plants. Total length rarely exceeds 6 inches. Indeed, if there is a seaweed that embodies the tenacity and stubbornness of northern coast people, it is Irish moss. It grows in such a way that it can survive almost all seasonal conditions. As a species, it embodies self-sufficiency, being a food and medicine all in one. In the northeastern United States, Irish moss is commonly found growing alongside a morphologically similar red algae, *Mastocarpus stellatus*, which is also edible and can be used for similar medicinal applications.

When torn from its disc-shaped holdfasts, the seaweed is bleached from dark reddish brown to almost white, washing up in wrack lines like some unusual fish cartilage. The life cycle of Irish moss is rather complex, alternating generations of gametophytes (male and female, producing half sets of chromosomes for reproduction) and sporophytes (asexual, containing a full set of chromosomes). Gametophytes and sporophytes can all be found growing in the same zone and are indistinguishable except for when the spores form. Irish moss can also reproduce vegetatively by growing new fronds from existing holdfasts or from any part of the thallus.

IRISH MOSS

MASTOCARPUS

Irish moss has smooth fronds, while mastocarpus has bumpy nodules found on the frond tips.

In upper tidal pools where there is more exposure to air and sunlight, the fronds of Irish moss shift from the characteristic deep purple to golden brown.

Carrageenan: Gooey Gold

The most distinctive feature of Irish moss, and the source of most of its medicinal benefits, is the presence of polysaccharides called carrageenan. Carrageenan is extracted from seaweed by heating it in water, upon which it forms a luscious, gooey gel. The Latin name *chondrus* comes from the Greek *khóndros*, meaning "cartilage," alluding to this connective tissue's flexibile qualities, which are also present in the bending movement of seaweeds containing gelatinous carrageenan.

Three Types of Carrageenan
All red seaweeds contain carrageenan in varying amounts, and there are actually three types of carrageenan—iota, kappa, and lambda—and Irish moss contains all three. Each of these types is differentiated by molecular structure and, as a result, the kind of gel it forms. Iota forms a soft gel (like well-set custard or jelly) and is frequently used in pharmaceuticals and cosmetics. Kappa forms a stiff and brittle gel (think hard-gel manicured nails) unless exposed to potassium salts, which creates more elasticity—it is often used in conjunction with locust bean gum to stabilize dairy products. Lambda, unlike the other types, does not form a gel but instead acts as a thickening agent used often to create creamy texture in dairy and dairy-alternative products. Irish moss's carrageenan is the most lambda heavy of all the red seaweeds. All carrageenans are soluble in water (iota and kappa require hot water), and the ratio of water to seaweed determines gel thickness. The gel can be firm enough to set a liquid in the same way that gelatin or agar (also extracted from seaweeds) does. This property is why Irish moss has been used as a food thickener for centuries.

The Carrageenan Boom
Irish moss was the base of an entire industry at one point, centered around the rocky coasts of Scituate, Massachusetts. In the mid-1800s, hundreds of Irish immigrants arrived in Massachusetts, fleeing the potato famine. They found opportunity in harvesting this seaweed, a familiar sight from their home shores. "Mossers," as they were called, and their families collected and dried the seaweed, then sold it to various manufacturers right as the carrageenan boom of the turn of the twentieth century hit. The mossers continued to prosper through the mid-1900s as World War II cut off the supply of agar (a similar seaweed compound) from Japan. As in most manual industries, however, the rise of mechanized production and the economic pressure of cheaper labor overseas forced the price of domestic hand-harvested Irish moss to rapidly rise. By 1995, the wild-harvest mossing industry had collapsed. Today, farms in the Philippines and Southeast Asia are the primary global suppliers of carrageenan.

MEDICINAL PROPERTIES

Energetics: Neutral to cool, moistening

Taste: The whole plant is salty, minerally, and a little bitter. The extracted carrageenan on its own still has some of that salty flavor but is largely tasteless and will take on the flavors it's prepared with.

Herbal Actions: Anti-inflammatory, antioxidant, antiviral, demulcent, emollient, hypoglycemic, hypolipidemic, nutritive, tonic, bulking laxative (mild)

Safety: There is no clinical research on the safety of Irish moss in pregnancy or lactation, but given that this plant has been used as a food for hundreds of years, it is most likely safe when consumed in culinary doses. Use caution when consuming the gel close to taking medications, as the seaweed mucilage may inhibit absorption.

Nutritional Value

This seaweed has a long-documented relationship with humans. It was a lifesaving food for coastal communities during the Irish potato famine of 1846 to 1848 and was eaten well before and long after those years. Seaweeds in general are excellent at taking in the dissolved vitamins and minerals in ocean water and storing them in their fronds, and Irish moss contains notably high levels of calcium, iron, iodine, potassium, magnesium, manganese, and copper. Irish moss is also high in protein; contains appreciable levels of the amino acids arginine, glutamic acid, citrulline, and taurine; and is an excellent source of polyunsaturated fatty acids (PUFAs).

Topical Emollient

Carrageenan gel is fabulous when applied to the skin! A cream that includes a carrageenan-rich seaweed infusion is great on burns (including sunburns), rashes, and scrapes. The mucilage helps keep moisture on skin longer and feels so silky. The gel can also be helpful in easing the itching and pain of psoriasis and eczema.

GI Support

Internally, carrageenan gel is wonderfully soothing to an irritated GI tract caused by ulcers, heartburn, sore throat, or IBS symptoms. It also functions as a prebiotic, feeding the favorable gut bacteria. During convalescence or postantibiotic use, it can be helpful when added to broth and soups. When taken with plenty of water, the soluble fiber of carrageenan also becomes a gentle bulking laxative that can help support bowel motility in periods of constipation.

Respiratory Support

In general, herbs and plants that are demulcent (or moistening) are considered soothing for the respiratory tract because of shared nerve channels between the nerves in the GI tract and those in the lungs. The result is that even though the plant constituents are not actually touching lung tissue, they have a "reflexive" soothing and tonifying effect on lung mucus, cilia, and smooth muscle. In the context of Irish moss, there are a lot of historical records stating its use as a therapeutic for coughs, bronchitis, and other lower respiratory conditions, especially on the cold, damp coasts of northern Europe. When Irish moss is cooked, the carrageenan gel that is so lovely on the skin and in the GI tract also becomes a remedy for respiratory issues. When the cilia in the lungs function properly, they can clear excess mucus up and out of the lungs, hence the use of Irish moss as an expectorant. Soothing the respiratory muscles helps reduce inflammation of the airways, making this a helpful tonic for asthma, COPD, and even long-term smoking or smoke exposure.

Cholesterol and Blood Glucose

In a small clinical trial, participants who received a carrageenan supplement for 20 days had markedly lower triglycerides and lower low-density lipoproteins (LDLs) at the end of the study. Another clinical trial had participants taking a carrageenan jelly or placebo for 60 days, after which those taking the jelly had significantly lower total cholesterol and lower LDL levels. Neither of these studies used whole-plant preparations, but given the ease with which carrageenan can be extracted from Irish moss, this seems like a place for so much more research and experimentation.

Supplementary Antiviral Support

Irish moss is one of the best-studied seaweeds for antiviral applications and has many potential uses as an adjunct antiviral therapeutic, meaning used in conjunction with other allopathic treatments.

For HPV: A 2019 clinical trial tested primarily the effectiveness of a carrageenan-based cream, Carraguard, against HIV infections, and secondarily high-risk human papilloma virus (HR-HPV) infections. While the cream was not effective against HIV, it significantly reduced the prevalence of HR-HPV in trial participants. Following this initial result, another clinical trial administered a carrageenan-based vaginal microbicide cream to cis women with genital HPV infections of various strains and resulted in 60 percent of patients testing HPV-negative at the conclusion of the study. Herbalist Lauren Giambrone has been working with HPV for over a decade and uses Irish moss tincture in her protocols alongside other antiviral and tissue tonic herbs, including astragalus, turkey tail mushroom, eleuthero, reishi, calendula, and licorice. She recommends mixing the infusion of Irish moss with chaparral in a sitz bath for active HPV infections and is also experimenting with making personal lubricants using carrageenan gel.

> **KEY TERMS**
> ## For Understanding Irish Moss Medicinal Properties
>
> **Cilia:** Fine hairlike structures in the lungs that move mucus and foreign particles up and out of the respiratory tract.
>
> **Endothelial and epithelial cells:** Endothelial cells line the blood vessels and lymphatic pathways, while epithelial cells are part of our skin and of the lining of the respiratory and digestive tracts.
>
> **Low-density lipoproteins (LDLs) and high-density lipoproteins (HDLs):** Molecules that function to carry cholesterol around the body. Generally, we want lower LDL and higher HDL levels.
>
> **Triglycerides:** A type of lipid (fat) found in the bloodstream.

For COVID-19: As we now know, the SARS-COV2 virus infects a body via the ACE-2 receptors that are present on cell walls of various tissues throughout the human body. Thus, a substance that inhibits ACE-2 binding effectively inhibits COVID-19 infection. Several in vitro studies have shown that sulfated polysaccharides from red seaweeds (carrageenans) competitively bind with the ACE-2 receptors, effectively blocking the spike proteins on the viral envelope of various coronaviruses from attaching and infecting the host cell. Two in vitro studies early in the pandemic focused on the application of carrageenan in nasal and mouth sprays to treat SARS-COV2. Human endothelial cell cultures (those cells that line blood vessels) were grown and administered carrageenan sprays, which showed effective inhibition of viral replication, without damaging the cells. After these studies, several clinical trials were conducted with human subjects, including one where 394 healthcare personnel were administered a carrageenan-derived nasal spray or placebo, and the carrageenan group had significantly lower infection rates. Several carrageenan nasal sprays are now available over the counter and are a highly recommended preventive measure against COVID-19 and other airborne viruses.

MEDICINAL PREPARATIONS

Irish moss is a perfect seaweed to begin experimenting with both because it is easy to incorporate into food and medicine and because the gooey carrageenan creates a sensory experience that brings you into the heart of the world of seaweed.

Tincture

A low alcohol percentage is best to keep polysaccharides intact, and the tincture should be gently heated to help carrageenan extraction. Lambda-carrageenan is soluble at the lowest temperatures, so of the red seaweeds, Irish moss is the best bet for a potent carrageenan-rich tincture. You can also use the double-extraction method, similar to making tinctures with mushrooms (see Appendix 1 for instructions).

Infusion

An infusion simply refers to a preparation of herbs steeped in hot or cold water. For seaweeds, and especially those containing carrageenan, we use a hot infusion.

To make an infusion of Irish moss:

1. Pour 2 cups boiling water over several fronds of seaweed.

2. Let sit for 15 to 20 minutes.

3. Strain out seaweed through a mesh or muslin bag so that you can squeeze the seaweed and let the goopy carrageenan plop back into your liquid.

This is a delightful sensory experience and a little different than a standard loose-leaf tea infusion that you may be accustomed to!

Tincture: Dried plant, 1:5, 40%, 1–3 mL per day

Infused Vinegar or Oxymel: Use fresh or dried whole fronds or flakes.

Infusion: Add flakes or whole fronds to teas and baths, or extract gel from infusion and add to other preparations.

Baths

Irish moss added to a hot bath is a luxurious experience. The fronds unfurl beautifully, and as the carrageenan seeps out, you can apply it to your skin and face. Combine with roses and goldenrod for a full seaside thalassotherapy (therapy by salt water) experience! If you don't have a bathtub, do a foot or hand bath instead.

To use Irish moss as a targeted perineal remedy:

1. Make an infusion with 2 cups boiling water and a few fronds of Irish moss.

2. Let sit for 15 to 20 minutes.

3. Squeeze seaweed a bit to get more goo out.

4. Strain and add to a sitz bath containing plain warm water or herbal infusion.

Cough Syrups

Irish moss has been a fantastic addition to herbal cough syrups for hundreds of years. As there are so many wonderful herbs for the respiratory system, I've differentiated these recipes into one for dry coughs featuring antitussive (cough-suppressive) and soothing herbs, and one for wet coughs featuring expectorant and drying herbs (see pages 56–57).

Carrageenan gel squeezed from Irish moss and prepared in a hot water infusion can be added to a bath or used in another herbal medicine preparation. Just a little seaweed makes a lot of gel!

This seaweed foot bath with Irish moss, calendula, yarrow, and roses is soothing to the skin.

HERBAL PARTNER FOR SITZ BATHS
Comfrey

Symphytum officinale is a plant whose giant leaves and enormous perennial roots may take over the corner of your garden, but its medicine makes it worth it. The compound allantoin in the leaves of comfrey stimulates epithelial cell growth, which means it closes wounds quickly. (Do not use it on open wounds that have not healed enough to be past the stage of potential infection.) It is also a first-choice remedy for bruises, sprains, and strains. It is particularly indicated in a postpartum sitz bath when the tissue may have slight tears.

Sitz Bath Infusion for
Inflamed or Damaged Tissue

MAKES 2 CUPS HERBAL INFUSION

Whether your skin is irritated from sweaty summer days or recovering postpartum, this infusion of skin-healing and soothing herbs can support you. The combination of yarrow and rose smells lovely, but you could also use a pinch of lavender if you prefer that aroma.

INGREDIENTS

- 1 part dried or 2 parts fresh Irish moss (2–4 fronds)
- 1 part calendula flowers
- 1 part yarrow
- 1 part plantain leaf
- ½ part comfrey leaf
- ½ part roses
- 2 cups hot water

INSTRUCTIONS

1. Prepare Irish moss infusion separately as described under Infusion on page 52.

2. Steep 3 to 4 tablespoons herbs in 2 cups hot water for 30 minutes. Strain before adding to total water volume used in your sitz bath.

Sitz Bath Infusion for Antiviral
and Antimicrobial Support

MAKES 2 CUPS HERBAL INFUSION

This blend can be used to help heal vaginal, penile, or perineal infections, alongside other pharmaceutical or herbal treatments. If you do not have access to chaparral, simply add a little more uva ursi. (This blend also makes a great foot bath for fungal infections.)

INGREDIENTS

- 1 part dried or 2 parts fresh Irish moss (2–4 fronds)
- 1 part calendula flowers
- 1 part lemon balm
- ½ part chaparral or uva ursi (whichever you have access to)
- 2 cups hot water

INSTRUCTIONS

1. Prepare Irish moss infusion separately as described under Infusion on page 52.

2. Steep 2 to 4 tablespoons herbs in 2 cups hot water for 30 minutes. Strain before adding to total water volume used in your sitz bath.

Cough Syrup for Dry Coughs

Dry coughs are often tight, barking, or raspy with little to no phlegm or mucus. The herbs in this syrup help to moisten and soothe the respiratory tract so that congestion can move and tissues can heal.

INGREDIENTS

- ⅓ cup dried wild cherry bark
- 5 cups cold water
- 2–3 fronds, or 2 tablespoons flaked, dried Irish moss
- 2 tablespoons dried hibiscus calyx
- 2 tablespoons dried licorice root
- 1 tablespoon grated fresh ginger
- 1 large cinnamon stick
- 1 cup raw honey
  Splash of brandy for extra preservative (optional)

INSTRUCTIONS

1. Add wild cherry bark to a medium-sized pot with 1 cup of cold water and let sit for 2 hours minimum, but ideally between 6 and 12 hours.

2. While that steeps, place Irish moss in bowl, cover with 2 cups of cold water, and let sit.

3. Drain Irish moss.

4. Add hibiscus, licorice, ginger, and cinnamon to medium-sized pot with remaining 2 cups water and bring to a simmer. Cook until liquid volume has reduced by half, 45 minutes.

5. Drain Irish moss and add to pot along with cherry bark and its water.

6. Continue to simmer for another 15 minutes.

7. Remove from heat and let cool for 15 to 20 minutes. Then strain through a strainer with cheesecloth or muslin. You should have about 1½ cups liquid.

8. Adjust amount of honey so it's proportional to the volume of liquid you have. Add honey to decocted (simmered and reduced) liquid and stir to dissolve. (Heat gently if necessary.) Add brandy, if using.

9. Decant syrup into clean jar and label. Syrup keeps in fridge for up to 3 months. Take 1 teaspoon as needed for cough, up to 6 teaspoons per day.

Cough Syrup for Wet Coughs

Wet coughs are often deep or gravelly, with copious phlegm or mucus. The herbs in this blend support a healthy cough that moves that congestion up and out of the respiratory tree.

INGREDIENTS

- ⅛ ounce dried Irish moss
- 6 cups cold water
- 1 ounce dried elderberries
- ¼ ounce dried elecampane root
- ⅛ ounce dried mullein leaf
- ⅓ ounce dried hyssop
- ⅛ ounce dried thyme
- 1 cup raw honey
- Splash of brandy for extra preservative (optional)

INSTRUCTIONS

1. Place Irish moss in bowl, cover with 2 cups of cold water, and let sit for at least 1 hour to rehydrate.

2. Drain and add to a medium-sized pot along with remaining 4 cups water and elderberries, elecampane, and mullein.

3. Bring to a simmer and cook, uncovered, for 30 minutes. Then add hyssop and thyme.

4. Continue to simmer until liquid volume has reduced by half, 20 to 30 minutes.

5. Remove from heat and strain through a strainer with cheesecloth or muslin. You should have about 1½ cups liquid (elderberries and roots absorb quite a bit of liquid that won't get pressed out, so you end up with slightly less than half of your starting liquid). Straining through a cloth is very important for this recipe, as mullein has tiny hairs that will be irritating to the throat if not strained out.

6. Add honey to decocted liquid and stir to dissolve. Add brandy, if using.

7. Decant syrup into clean jar and label. Syrup keeps in fridge for up to 3 months. Take 1 teaspoon as needed for cough, up to 6 teaspoons per day.

HERBAL PARTNERS FOR COUGH SYRUP

There are many herbs you can pair with seaweeds to enhance your syrup's soothing properties.

Wild Cherry Bark

Wild cherry (*Prunus serotina*) is one of the best antitussive (cough-reducing) herbs. Generally we use the bark, because the leaves and seeds of the fruit contain potentially harmful levels of cyanogenic glycosides. The bark's most distinctive feature is its strong almond-like flavor and aroma that is most pronounced when it is steeped in cold or room-temperature water. It's a cooling respiratory relaxant and antispasmodic, perfect for hot, dry, and unproductive coughs. The bark becomes more astringent when infused or cooked in hot water, which is why the recipe here limits the cooking time for the bark in the syrup.

Hibiscus

Roselle hibiscus (*Hibiscus sabdariffa*) is in the mallow (Malvaceae) family and carries both demulcent (soothing) and astringent (tonifying/tightening) properties. We use the calyxes of the flowers, which turn any liquid bright reddish pink. They are sour and refreshing, and they're packed with a healthy dose of vitamin C. Hibiscus is sometimes referred to as "sorrel" (not to be confused with garden sorrel in the Polygonaceae family) and "Jamaican sorrel"—while the plant is native to West Africa, it was brought to the Caribbean via enslaved peoples. If you don't have access to dried roselles, you can substitute rose hips for the vitamin C boost.

Licorice Root

Licorice (*Glycyrrhiza glabra*) doesn't actually taste like licorice candy, which is flavored with anise seed. The flavor of this root is a bit earthy and delightfully sweet, with just a whisper of anise. This plant is used in Traditional Chinese Medicine (TCM) and is native to West Asia, North Africa, and Europe, but it grows well in temperate conditions in the US. Licorice is a fantastic anti-inflammatory with specific indications for respiratory and digestive tracts. It is both a mild expectorant and soothing for sore throats. In TCM formulas, licorice is often used in small amounts as a "harmonizer," bringing all the components together. Licorice taken frequently in large amounts can raise blood pressure, so omit it from the recipe or do not use longer than a couple of weeks if you are on blood pressure medication or have hypertension.

Elecampane

Elecampane (*Inula helenium*) is a gorgeous towering plant with an enormous root structure. We use the roots as a powerful expectorant to promote a productive cough. Elecampane has a deep, bitter, earthy flavor and is indicated for clearing crud from the depths of your lungs. It's a great remedy for acute illness and for chronic respiratory conditions alike. A little goes a long way, but when combined with honey (like in Cough Syrup for Wet Coughs, page 57), it becomes quite complexly delicious.

Hyssop

Hyssopus officinalis is a plant many people will recognize the flavor and smell of because it is one of the primary ingredients in Ricola cough drops. It's minty and resinous, like a combination of sage and peppermint. Hyssop is both a stimulating and relaxing expectorant, making it a great remedy for damp congestion and a tired respiratory tract that has been working really hard to clear the gunk. A hot infusion of this plant is also a great diaphoretic—it promotes sweating and helps break a fever.

Mullein

Mullein (*Verbascum thapsus*) is a tried-and-true respiratory tonic. The large lobed leaves resemble lungs and are a classic example of the Doctrine of Signatures—the theory that a plant will look like the part of the body for which it is a remedy. Indeed, it is a respiratory antispasmodic and pectoral tonic (for chest and lung muscles). It is usually indicated for dry coughs, but I find that it's a good balancing element in the formula for wet coughs on page 57. The tall, yellow-flowered spires of these biennial plants are hard to miss in fields and along roadsides. Mullein leaves are so soft because they are covered with tiny hairs, which can be irritating to the throat if you don't use a fine strainer for infusions and other liquid preparations.

Elderberry

Sambucus spp. can be found growing near waterways and damp places, with gorgeous showy umbels of white flowers in spring that turn into bunches of deep blue-purple berries. Elder flowers are a great diaphoretic for breaking a fever, and the fruits are strongly antibacterial and antiviral, helping boost the immune system during acute illness. The berries are exceptionally high in anthocyanins, giving them excellent antioxidant potential. They also taste delicious—but cook or steep them in tea first to avoid any stomach upset.

Blancmange Pudding

In general, you can use powdered Irish moss as a thickener in soups and stews, as the hot water will transform the carrageenan into a gel. You can also add it to smoothies, but I would recommend *Gracilaria* (Jamaican sea moss) for this purpose. One of my favorite recipes with Irish moss is a traditional pudding that comes from Ireland. Sweetened with honey or sugar and flavored with vanilla, chocolate, or fruit, you'll be hard pressed to even detect the seaweed. I've served this to a class of middle schoolers who generally agreed that it was tasty!

MATERIALS

Double boiler (if using dairy milk)
Muslin bag or cheesecloth

INGREDIENTS

½ cup Irish moss flakes or whole leaf broken into pieces

4 cups milk of your choice (cow or almond/coconut/oat)

¼ cup sugar or ½ cup honey (or to taste)

Flavoring of your choice: Options could include cacao powder, vanilla extract, lemon zest, macerated raspberries or strawberries, cardamom pods, ground cinnamon, or ginger. I am partial to adding ¼ cup or so of cacao powder and some vanilla extract.

INSTRUCTIONS

1. Rinse seaweed several times with cold water to remove any rocks, shells, or excess salt deposits.

2. Soak seaweed in a bowl of cold water for about 30 minutes. (This rehydrates seaweed and helps remove some of the briny flavor.)

3. While seaweed is soaking, pour milk into the top half of a double boiler and begin to warm. (If you are using nondairy milk, you do not need a double boiler, as there is not the risk of burning it in the same way.)

4. Put seaweed in a muslin bag (or cheesecloth, tie closed) and add to milk.

5. Simmer seaweed in milk for 30 minutes to 1 hour, stirring occasionally and squishing the seaweed bag against the side of the pot with a spoon to squeeze out carrageenan.

6. Add sugar or honey and your flavorings of choice after about 20 minutes and continue stirring until fully incorporated. (These need to be added before there is enough carrageenan in solution to start gelling.)

7. Once the mixture starts to noticeably thicken, turn off heat and pour into molds, cups, or whatever container you'd like to serve from.

8. Cool on the counter for 20 minutes, then in the fridge for at least 1 hour, up to overnight, until set. Add any additional toppings and serve.

Atlantic Holdfast Seaweed Company

A tlantic Holdfast Seaweed Company is a one-man operation dedicated to sustainable wild-harvesting, bringing together founder Micah Woodcock's love of botany and horticulture, sustainable food production, and being immersed in the marine environment.

Micah is a lifelong Mainer who learned to love the ocean during the years of his childhood spent in Greece for his parents' work. When he returned to Maine, he became curious about the small industry of seaweed harvesters on the down-east coast, apprenticing in the art and science of seaweed harvest with Larch Hanson, an old-timer of the Maine seaweed scene. After one season, Micah was hooked. Now his company provides seaweed to individuals, herbalists, and seaweed enthusiasts looking to bring these gorgeous plants into their kitchens and medicine cabinets.

A Community Resource

During the seaweed season, Micah ventures into the water to harvest Atlantic kombu, Irish moss, bladderwrack, and sugar kelp. After 12 years, he has developed a deep awareness of the seasonal cycles of seaweeds that inhabit the wildly dynamic intertidal zones of Maine. Recently he's noticed a decline in *Alaria,* possibly due to out-competition from *Chondrus* and *Laminaria digitata,* whose spores are more resilient in rising seawater temperatures. He's calculated just how long the high intertidal bladderwrack lines spend out of the water versus submerged (60 to 70 percent of their day is out of the water, in case you're wondering!). This is the kind of attention to detail that successful and sustainable wildcrafting requires.

The folks who wild-harvest seaweeds on the Maine coast are a tight-knit community where everyone knows each other's harvesting zones, so they don't overstep or create unnecessary scarcity of a resource that is so abundant. Micah used to be more concerned about folks' enthusiasm for wild-harvesting seaweed. Now that mariculture is taking off, however, he is less worried about pressure on the wild stocks. He points out that mariculture has the potential to be sustainable even at large scale, while wildcrafting is not and never will be. While wildcrafting remains a small industry, he would like to see more conversation about wildcrafting and wild-tending of seaweeds as an integral part of sustainable local food systems, and not just a way to upcharge fine-dining dishes and culinary products.

An Ecosystem of Relationships

Micah's harvesting practices revolve around understanding relationships to ensure that resources are not overused or exploited. Generally, capital is the limiting factor in resource use—whoever can pay the most gets access. In the case of wildcrafting, however, those who have apprenticed to previous generations of harvesters and to the teachings of the sea and seaweeds themselves have access to the bounty of the intertidal.

Part of what keeps Micah on these cold, rocky shores is the joy of the different motions required to harvest each seaweed. Atlantic kombu is his favorite to harvest because it requires being fully immersed (in order to cut it with a knife), feeling the push and pull of the waves. It's very different from gathering Irish moss, which is collected either by hand or with mossing rakes while either balancing on ledges or harvesting from a boat. Micah loves Irish moss—he recommends not just eating it but also using the carrageenan gel as hair conditioner, detangler, and styling cream all in one!

Micah notes that seaweed wild-harvesters and farmers need to pay attention to the entire ecosystem, including the fish and invertebrates that inhabit and feed on the algae, particularly with regard to larger-scale harvesting of *Ascophyllum* for fertilizer and animal feed. The seaweed itself may be harvested sustainably, but how does harvesting impact other organisms? We don't fully know right now. Close observation and slow, intentional, and relational tending of these places will likely give us answers.

As with growing medicinal herbs, seaweed farming and wild-harvesting need to happen on different scales to be accessible and sustainable. Large farms have the capital to produce kelp on the scale required by the current food processing systems, but we also need small farms that support local economies and have the excitement to test new technologies, and we need the network of wild-harvesters who keep an eye on the places and species that others might overlook, and who are keepers of the seaweed-tending lineages that we cannot let ourselves forget.

Wildcrafting and tending wild seaweeds is an integral part of sustainable local food systems.

Botany & Ecology

Rockweed (*Ascophyllum nodosum*) is one of the predominant seaweeds found along the New England intertidal zone, and its range extends across the entirety of the northern Atlantic. In recent years, rockweed has made its way to northern Pacific shores and is considered invasive there because once established, it tends to remain and outcompete other species. It is also a bit unusual in that it can tolerate low salinity and so can be found up in estuaries where other seaweeds would not survive the brackish environment. This seaweed is so abundant in the intertidal that it's easy to ignore—like sweeping your eyes over a field of grasses that all more or less look like the same tall, swaying green forms. You have to get a little closer, run your hands through the fronds, and you'll start to notice their unique shapes.

Identifying Rockweed

Although a brown seaweed, rockweed looks distinctly olive green even when drying in the sun between tides. It grows on sheltered rocky surfaces in the mid-intertidal, attached with a disc-shaped holdfast. You can't mistake its spaghetti-like growth habit, with elongated leathery flattened fronds that can grow up to 6 feet. Branching from the central thallus is common, but branches will remain shorter than the primary stalk. Oval-shaped air bladders form at regular intervals along the fronds and give rockweed its distinct floating movement when submerged (and its common name: egg kelp). You can estimate the age of a given rockweed by counting the air bladders along that central thallus. Bladders generally won't form for the first 2 to 4 years, but after that, each season's growth adds one more bladder. Rockweed doesn't grow particularly fast (for a brown seaweed), but it is tenacious—its long growth habit allows for ends to get broken off in storms or by grazing by marine invertebrates without permanently damaging the individual. Rockweed has one of the most straightforward reproductive cycles, with dioecious plants (male and female gametes produced by separate individuals) that form spore receptacles containing egg or sperm gametes that are released into the water column in early summer.

A Culinary History

Rockweed has a long relationship with people. While not particularly palatable when freshly harvested due to its tough texture, this seaweed, along with other intertidal species, has long been used for cooking processes. Beachside clam and lobster bakes do not actually originate in the Americana they are enshrined in. Steaming shellfish and starches over hot coals covered in seaweed is an Indigenous culinary practice that European settlers took up and claimed as their own foodway. In any case, infusing the salty flavor of seaweed into proteins is something that humans have enjoyed for centuries. See page 72 for recipe suggestions to get you started.

This shallow, sandy coastline has a classic zonation of bladderwrack growing higher in the intertidal (bottom right) and rockweed inhabiting the lower areas (bottom left and upper right).

Ascophyllan: Potential for Cancer Support

Brown seaweeds contain a variety of phycocolloids (gelling compounds found in seaweeds), and rockweed has its own unique contribution: ascophyllan. Like the fucoidan of *Fucus* spp., ascophyllan is a polysaccharide with a fucose (a sugar) backbone. This similarity, combined with the relative abundance of *A. nodosum*, has led to increased interest in clinical applications of ascophyllan. While there are no published human clinical trials as of this writing, animal research has shown that ascophyllan extracts stimulate immune cells and inhibit cancer cell growth. A 2014 study showed that ascophyllan extract induced maturation in dendritic cells (the primary cells responsible for antigen-specific immune system response) and subsequently boosted immune system chemical signaling. An earlier study in 2009 demonstrated increased production of tumor-necrosis factor (TNF-alpha), which is generally associated with anticancer actions in plants. On the whole, the effects of ascophyllan as an antitumor agent are highly dependent on the particular cancer cell line, and more research is needed to tease out those particulars.

Phlorotannins: Versatile Compounds

A type of phenolic compound, the plant chemical phlorotannin is present only in brown seaweeds. Botanically, phlorotannins help protect against oxidation and predation from herbivores. Content in the thallus varies by location and by season (highest in summer) but can be up to almost 6 percent of dry weight in *Fucus* spp. and *A. nodosum*.

Many Potential Uses

Various studies have confirmed the bioactivity of phlorotannins to include antioxidant, anti-inflammatory, anticancer, antiviral/antimicrobial, hypotensive, neuroprotective, and prebiotic. Additionally, it appears the phlorotannins are significantly more bioavailable in the human gut than are many other compounds. Phlorotannin extracts from *A. nodosum* demonstrated antihypertensive (blood pressure lowering) effects via inhibiting glucose uptake and reduction of total cholesterol in diabetic mice. While human clinical trials on rockweed extracts are still inconclusive, these results closely follow those from similar experiments with fucoidans from *Fucus vesiculosus*.

A positive correlation between phlorotannins and anti-inflammatory and antioxidant actions in epithelial (skin) cells gives clinical basis to the long history of topical seaweed therapies. There is also some fascinating research in development looking at the neuroprotective effects of these phlorotannins extracted from *F. vesiculosus* and other brown seaweeds, from memory-enhancing action to circadian rhythm regulation by interaction with GABA receptors. While the mechanism remains unclear, there have also been preliminary trials showing potential mood regulation and antidepressant effects of these phytochemicals as well. Finally, tannins as a class of molecule are also highly effective as natural dyes and are the reason why bladderwrack has historically been called dyer's wrack. (See Appendix I, Natural Dyeing with Seaweed.)

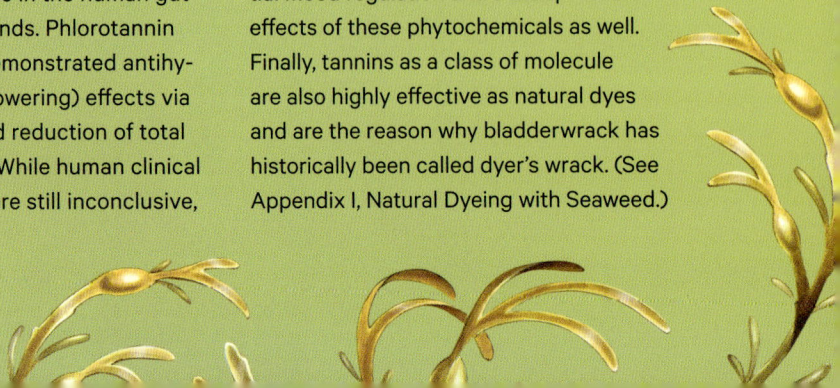

MEDICINAL PROPERTIES

Energetics: Cooling, moistening

Taste: Salty, minerally, slightly bitter, umami

Actions: Anticancer, anti-inflammatory, antioxidant, antiviral, diuretic, emollient, hypoglycemic, hypolipidemic, hypotensive, neuroprotective, nutritive, prebiotic, thyroid stimulant

Safety: There is no clinical research on the safety of rockweed in pregnancy or lactation, but it is most likely safe when consumed in culinary doses. Use caution with hyperthyroid conditions or when used in conjunction with hypothyroid medications. Potentially contra-indicated with blood clotting disorders and/or anticoagulant medications due to synergistic effects, however no clinical research has been done on these interactions.

Nutritional Value

Due to its leathery texture, rockweed is difficult to eat whole. Grinding it up, however, makes it easy to add to food or tea so you can access the nutrients. As a brown seaweed, it is high in iodine, which makes it a mild thyroid stimulant. Additionally, it is high in vitamins C and E, as well as iron, manganese, and magnesium. There is some protein, but not as much as many of the red or green seaweeds. Rockweed is also high in dietary fiber.

Metabolic Syndrome Support

Brown seaweeds, including rockweed, offer an array of support for managing (and preventing!) metabolic syndrome—a cluster of symptoms including high blood pressure and dysregulated blood sugar and cholesterol that increase risk of diabetes, heart attack, and stroke. Therapeutics for metabolic syndrome include medications, supplements, and dietary interventions that help to regulate blood sugar, protect vascula-ture, lower blood pressure, and balance choles-terol. Seaweed can support all of these goals!

THE ROLE OF FUCOIDANS

Fucoidans are polysaccharides in brown algae that support cardiovascular health (see also Winged Kelp and Wakame; Bladderwrack; and Sugar Kelp and Kombu Kelp) in a multi-tude of ways. Polysaccharides in rockweed are similar to those studied in reishi mushrooms (*Ganoderma lucidum*), which support the gut microbiome in the lower intestine, which ulti-mately helps manage cholesterol and blood sugar dysregulation by feeding the bacteria that are most efficient at cholesterol breakdown.

In the small intestine, enzymes including a-glucosidase and a-amylase quickly break down carbohydrates into simpler sugars for easy absorption into the bloodstream. Multiple antidiabetic medications utilize enzyme inhibitors to slow the action of these enzymes, prolonging digestive time and glucose absorp-tion, thereby regulating blood sugar. Fucoidan extracts from various seaweeds have been tested for similar effects, and extracts from *A. nodosum* are some of the strongest inhib-itors of a-glucosidase and a-amylase. In a variety of human clinical trials, patients have been administered capsules of *A. nodosum* or *A. nodosum* combined with *F. vesiculosus*, and the results have been inconclusive. Some trials show significant changes in triglycerides, insulin levels, and inflammatory markers; others show no difference from the placebo administration. In no instance has seaweed intake made these metrics worse.

We also know that the health benefits we see from *A. nodosum* (and *F. vesiculosus*) regarding metabolic syndrome are due not just to the well-studied fucoidans but also to the very high antioxidant activity and phenolic content of these seaweeds. With all this in mind, rockweed seems to bear out as an incredibly promising functional food, with only potential benefits to be gained from adding it to an herbal and/or dietary protocol for managing metabolic syndrome.

Thyroid Tonic

Rockweed, like all brown seaweeds, contains significant concentrations of iodine, which supports healthy thyroid function and can be a boost for hypothyroid conditions. The high iodine concentrations are part of what gives rockweed its distinct bitter flavor. If you are using rockweed for thyroid support, capsules are often the best method of delivery. (See Bladderwrack for further details.)

Adjunct Therapeutic for Cancer

Fucoidans (see Winged Kelp and Wakame) have been extensively studied for anticancer activity. Extracts from *A. nodosum* have been shown to halt cell replication and reduce chemo-resistance (unresponsiveness to chemotherapy treatments) in some human lung cancers.

Topical Antioxidant

Rockweed and bladderwrack have both been used together for skin health in Traditional Chinese Medicine and European traditions. At least one in vitro study has confirmed the antioxidant and anti-inflammatory benefits of *A. nodosum* phlorotannin extracts on epithelial (skin) tissue. For topical use, I would recommend a compress, a poultice, or even a liniment as the best delivery method. (See Appendix 1: Herbal Preparations 101 for how to make these preparations.)

KEY TERMS
For Understanding Rockweed Medicinal Properties

Angiotensin-converting-enzyme (ACE): A protein whose function is to help regulate blood pressure by increasing vasoconstriction, or the tightening of blood vessels. In healthy vasculature, this is a normal part of controlling the volume of fluids in the body and keeping blood pressure stable. In vasculature compromised by disease or inflammation, this constriction becomes potentially dangerous, elevating blood pressure too high or even blocking blood flow if substantial buildups of plaque (fats, cholesterol, etc.) are present. ACE-inhibitor drugs are generally effective and are commonly prescribed.

GABA: The primary inhibitory neurotransmitter in our central nervous system. GABA is a set of molecular brakes that has calming effects when activated.

MEDICINAL PREPARATIONS

While rockweed is not usually eaten whole, it's a great seaweed to infuse into broths and vinegars. Due to the high iodine content, it is also a good choice for powdering and making into capsules for thyroid support.

Tincture: Dried plant, 1:5, 30–40%, 1–3 mL per day

Capsule: Powdered whole thallus, 0.5–2 g per day

Infused Vinegar or Oxymel: Use fresh or dried, whole thallus or flakes.

Infusion: Add flakes to a tea blend.

HERBAL PARTNERS FOR ROCKWEED

The following are great food-as-medicine partners used in combination with rockweed.

Nettle

Stinging nettles (*Urtica dioica*) are a powerhouse of nutrients. Often compared with spinach, they contain high levels of iron, zinc, copper, manganese, nickel, and vitamins C, B2, and K. Among their many uses is as a nutritive, fortifying tonic. Their mineral-y flavor also plays exceptionally well with mushrooms and seaweed.

Astragalus

Astragalus membranaceus comes from Traditional Chinese Medicine (TCM), where it is considered a lung and qi tonic. It's adaptogenic and slightly immune-stimulating, which makes it a great choice for broths (and slow-cooker beans!) when the weather turns cooler and folks start getting sick more often. The part of the plant used is the root, which is often sold in thin, tongue depressor–shaped slices. Note: Do not use astragalus when actively sick; it functions much better as a preventive.

Codonopsis

Codonopsis pilosula, or bell-flower, is another TCM herb that is absolutely wonderful for broths, soups, or anything simmering in a pot. Codonopsis is an adaptogen and digestive tonic that is especially helpful in times of fatigue or weakness post-illness. The easily purchased slices cook down quickly and taste a bit like a sweet and earthy carrot.

Tidepool Beans

Seaweeds absolutely shine when added to beans. Seaweed adds salt to the beans and minerals to the broth, and it contains enzymes that reduce the gassiness of beans in your intestines! Any kind of seaweed works, but there is something so amusing and satisfying about watching whole pieces of rockweed drift around the pot as the beans simmer. This recipe comes from a former partner who is both an herbalist and a great cook, making dishes that sit squarely in the food-as-medicine category. Their favorite herb and spice additions include medicinal mushrooms, codonopsis, astragalus, nettle, and hibiscus or sumac.

INGREDIENTS

- 1 cup dried beans of your choice
- 2–5 dried shiitake, oyster, or lion's mane mushrooms
- 1 (or two) 6-inch pieces of dried or fresh rockweed
- 1 whole dried chile of your choice (or more if you want more spice!)
- ¼ cup dried nettle
- 1 tablespoon dried codonopsis rondelles
- 2–5 pieces astragalus
- 1 bay leaf
- ½ teaspoon ground coriander
- ½ teaspoon ground cumin
- ½ teaspoon dried thyme
- ½ teaspoon dried oregano and/or bee balm
- ½ teaspoon dried sumac powder (optional for a nice tart note; you could also add a splash of apple cider vinegar)
- ¼ teaspoon freshly ground black pepper
- Salt

INSTRUCTIONS

1. Rinse dried beans in a strainer. Put them in a slow cooker or large pot, cover with ample water, and let soak for 1 to 12 hours. Strain.

2. Chop mushrooms into pieces if you'd like to eat them, or keep whole if you want to eat around them.

3. Add enough water to the pot to cover beans by 2 inches, then add mushrooms, seaweed, chile, nettle, codonopsis, astragalus, bay leaf, coriander, cumin, thyme, oregano, sumac, black pepper, and salt to taste.

4. Bring to a boil, then simmer with the lid on until beans are soft, 2 to 4 hours, adding more water as needed to keep beans submerged and prevent them from sticking while simmering. Cooking time will vary slightly depending on the variety of bean.

5. Once beans are done to your liking, remove rockweed, astragalus, bay leaf (and whole mushrooms, if desired).

Clambake . . . on Your Stove!

SERVES 4

This chapter wouldn't be complete without including a variation on the clambake.
Most of us don't have access to a beach to build a full pit fire for a traditional bake.
You can create the flavors and aromas in your kitchen, however, with just a large
pot and a steamer basket. You can use any variety of fresh clam available, as well
as additional seafood (crab, lobster, fish fillets) or sausage to make this a heartier
meal. I am partial to the classic corn and potatoes, but you can get creative with
veggies, too. Don't forget the herbs and lemon!

INGREDIENTS

- 5 teaspoons salt
- 4 cups cold water (below 65°F/18°C)
- 2 pounds fresh clams (steamers, little neck, cherrystones, or countnecks)
- 1 ounce dried rockweed, or a large handful of fresh rockweed (enough to line the bottom of the steamer basket)
- 1 bunch fresh thyme
- Several sprigs fresh rosemary
- 2 white onions, peeled and quartered
- 2 ears of corn, cut in half
- ½ pound red or yellow potatoes, cut in halves
- 1 lemon, sliced into thin rounds

FOR SERVING

- 2 scallions
- 2–4 cloves garlic
- 4 tablespoons butter
- Freshly cracked black pepper
- Additional lemon slices or lemon juice

INSTRUCTIONS

1. First clean the clams: Mix salt and cold water. Place clams in water and let sit in the fridge or on the counter on ice for 2 hours. The water will end up with gunk and sand that is purged from the clams' siphons.

2. Before starting to cook, make sure all clams are alive. When tapped gently, they should close their shells tightly. Discard any dead clams.

3. Prepare seaweed: If using dried rockweed, put in a bowl and cover with cool water. Let sit for 15 to 30 minutes to rehydrate. If using fresh seaweed, just rinse any shells or sand off fronds.

4. Fill a large pot with water up to the bottom of a steamer basket set inside the pot.

5. Drain seaweed and spread in a layer across the bottom of the steamer basket.

6. Add thyme and rosemary on top of seaweed, then add onions, corn, and potatoes.

7. Place clams on top of vegetables, and tuck lemon slices between them.

8. Bring water in the pot to a boil and cook with lid tightly secured for 30 minutes.

9. While clams steam, make butter sauce: Sauté scallions and garlic in 2 tablespoons of butter until fragrant, 5 to 7 minutes. Add pepper and sauté for an additional 2 minutes. Remove from heat and add remaining 2 tablespoons butter and allow to melt fully.

10. Once all clams have opened, your clambake is done! Serve immediately with butter sauce drizzled over the top and additional lemon.

Seaweed Infused Chocolates

FINAL QUANTITY OF INDIVIDUAL CHOCOLATES
VARIES DEPENDING ON THE SIZE OF YOUR MOLD

SERVES 4–6

Ascophyllum nodosum (rockweed) contains higher levels of vitamin E than red and green seaweeds, which we can take advantage of by combining it with chocolate. Vitamin E is lipid (or fat) soluble, so while this recipe doesn't involve a long enough infusion to extract the vitamin E, the fat contained in chocolate makes the vitamin E in the seaweed more bioavailable. Making herb-infused chocolates is actually very easy, and I've found that adults and kids alike enjoy the process and the result. For this recipe, you'll need silicone molds to pour the melted chocolate into. You can find all sorts of them at craft stores, baking stores, or online. Adjust the amount of ingredients depending on the volume of your molds. You can get creative with flavors and add other herbs or spices as you like! If you use an herb powder, mix it in with the seaweed. Otherwise, just add it to the bottom of the molds.

INGREDIENTS

- 1 tablespoon dried rose petals, ground
- 1–2 tablespoons finely chopped dried fruit, such as cape gooseberries or candied ginger (optional)
- 20 ounces (two bags) dark chocolate chips (or milk chocolate if you prefer)
- 1 teaspoon dried and powdered rockweed (you can grind it yourself in a spice grinder)

INSTRUCTIONS

1. Set the candy molds on the counter and add a sprinkling of rose petals and chopped dried fruit to the bottom of the molds.

2. Place chocolate chips in the top of a double boiler. Heat gently over boiling water and stir chips frequently to melt them.

3. Once chocolate is just melted, reduce the heat under the double boiler to a simmer and add seaweed powder. Continue stirring until well combined, approximately 5 minutes.

4. Remove the double boiler from the heat and use a spoon or pastry bag to transfer chocolate to molds.

5. Chill in the fridge for at least 1 hour before removing from the molds.

Sugar kelp, with its thick, long fronds, needs to live in water deep enough to remain submerged at all times.

Botany & Ecology

Although several sizes and morphological variations exist, sugar kelp is generally easy to recognize, with its clawlike holdfast; short, thick stipe; and long, broad, golden brown blade with frilled margins and no midrib. One form of sugar kelp has a hollow stipe. Another has fronds that are more streamlined and straplike due to growing in habitats with more wave action. Under ideal conditions, individuals can grow to more than 16 feet long.

Wild Growth

Sugar kelp is rather adaptable—it's found at the edge of the subtidal but also exists happily as deep as 100 feet. Their holdfasts are not too picky either, attaching to rocky ground, small rocks, and debris, or to lobster trap lines and other fishing gear. This is part of the reason that sugar kelp is so easy to farm on long lines set in the water column. As is true of all kelp, it grows remarkably fast (up to 2 inches per day) during an individual's 2- to 5-year lifespan. If you cut open the stipe, you can see thin rings of annual growth, just like you would see in a cross section of a tree trunk. During the reproductive season (remember, the kelp fronds that we see are the sporophyte phase), the blade develops a dark band of tissue, which releases spores.

A Keystone Species

Sugar kelp is the predominant kelp species being farmed on the Northeast coast of the US, but its range is throughout the North Atlantic and the Pacific. The plants survive just fine under ice but can only tolerate water as warm as 75°F (24°C). Long Island Sound and Northern California are the current southern extents of its range, but that latitude threshold continues to shift farther north as ocean temperatures rise. Ecologically, sugar kelp is as much of a keystone species as giant kelp, providing enormous amounts of habitat, food, and shelter for other marine organisms.

The Kelp Highway theory posits that human migration along coastal routes was possible because of kelp forests. People traveled across the Bering Strait and down the coast of North America, following the kelp. Kelp forests stop around the equator but continue again on the Pacific coast of South America. Kelp has long been a food in and of itself, but it's also a key habitat for shellfish, mollusks, and fish that people rely on.

In addition to being eaten, kelp has been, and continues to be, used around the world as raw material in everyday households and industry alike. Kelp stipes were used for weaving and as living moorings for boats. Fast-forward to the eighteenth century, and kelp became a major industry on many coasts, with much documentation in Scotland and the British Isles. Kelp was burned with rockweed and bladderwrack to make iodine, as well as potash (potassium carbonate) and soda ash (sodium carbonate)—indispensable for the glass and soap industries.

Ecklonia
Laminaria
Macrocystis
Saccharina
Lessonia
Nereocystis
Eualaria
Human migration path

The Kelp Highway theory posits that coastal human migration from eastern Asia, across the Bering Sea, and down the Pacific coasts of North and South America was fueled by the resources of kelp forests.

Today, most commercially farmed sugar kelp is used for food (for both humans and animals) and for the extraction of alginates.

Medicinal and So Much More: Alginate Gel

Alginates are another medicinal and widely utilized seaweed gel. While all brown seaweeds contain alginates, kelps have the most (30 to 45 percent dry weight) and are the primary source of alginates worldwide.

Alginate refers to the substance formed when the alginic acid (algin) combines with sodium or calcium, forming a hydrocolloid polysaccharide, which is a chemistry term for a type of gel. This gelling property was recognized and used as early as the classical period (eighth century BCE to fifth century CE) to fireproof ships in the Mediterranean. Alginates were officially classified at the turn of the nineteenth century and were quickly applied across industries, including continued use as a fire retardant in firefighters' uniforms. Sodium alginate is water soluble and is used today primarily as a stabilizer and emulsifier in dairy products and ink. Calcium alginate is insoluble in water and is used in textiles and bandages and in the cosmetics industry.

Like other polysaccharides, algin is present in the cell walls of seaweeds, helping give them flexibility and motion in the waves. Each species of kelp produces a slightly different version of algin, which makes harvest location, time, and season very important for achieving desired consistent extracts.

SEAWEED SCIENCE
Fucoxanthin: An Accessory Pigment

Fucoxanthin belongs to a class of molecules called carotenoids. Carotenoids are very familiar to many of us, giving carrots, squash, melons, sweet potatoes, and citrus their characteristic orange color. Beta-carotene and lycopene are two of the most common forms that we ingest. In algae, fucoxanthin is found in chloroplasts and acts as an accessory pigment, improving light harvesting and energy transfer to the primary pigments (chlorophyll). Macroalgae and microalgae alike produce fucoxanthin, which provides a brownish yellow color to the organism. There is some uncertainty about the bioavailability of fucoxanthin, but clinical research suggests that when whole seaweed is ingested, the synergistic effects of other compounds make the therapeutic benefits of this pigment more available to our bodies.

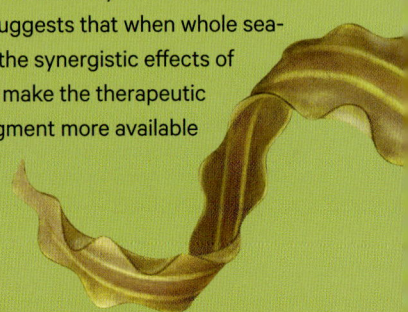

Sweet

Sweet is one of the five tastes (alongside salty, sour, bitter, and umami) that our tongues perceive, though it's not one usually associated with seaweed.

The Many Types of Sweetness

Our cultural understanding of "sweet" has shifted dramatically since the industrialization and global trade of processed cane sugar. Cane sugar, jaggery, honey, maple syrup, and other classic sweeteners are undeniably sweet, due to the concentrated fructose, glucose, or sucrose contained in each. However, there are many types of sweet that exist in edible plants and herbs, including the fresh, tangy notes of summer berries, the starchy caramel of a roasted sweet potato, and the almost synthetic-tasting flavor of stevia leaves.

In all energetics systems, sweet connotes nourishment. Our bodies know that sweetness means carbohydrates (which break down into glucose to feed cellular metabolism), which equals energy and vitality. Our mouths, by and large, love sweetness. Our tissues soften in its presence—"a spoonful of sugar helps the medicine go down" is really true! This is one of the reasons why the herb licorice (*Glycyrrhiza glabra*) is considered a harmonizing herb in Traditional Chinese Medicine. The sweet taste of the root helps blend other stronger herb flavors together and encourages the body to more readily receive the medicine of the whole formula. The roots of many tonic herbs are at least slightly sweet, such as codonopsis (*Codonopsis pilosula*), platycodon (*Platycodon grandiflorus*), burdock (*Arctium lappa*), marshmallow (*Althea officinalis*), and even early-season dandelion (*Taraxicum officinale*). In these plants, the line between food and medicine becomes blurred.

Sweetness from Seaweeds

So what about seaweed? Due to their growing environment and nutrient-concentrating properties, seaweeds are understandably perceived as salty. They have an inherent sweetness, however, that comes from the starches and laminarins (starchlike molecules in brown seaweeds) that store carbohydrates (glucose) for the plant. Sugar kelp gets its name from the naturally occurring mannitol sugars within the plant; they crystallize on the surface, along with mineral salts of sodium and potassium, as the fronds dry.

Mannitol has 60 percent of the relative sweetness of sucrose and is not readily absorbed by the human GI tract, so you may see it used as an alternative sweetener in a variety of food products. (For comparison, coconut sugar has about 75 percent of the relative sweetness of table sugar/sucrose.) Sugar kelp also contains high levels of the amino acids alanine and proline, which taste sweet.

This dried kelp has a layer of mannitol and salt on the surface—not to be confused with mold.

MEDICINAL PROPERTIES

Energetics: Neutral to cool, moistening

Taste: Salty, mineral-y, mildly sweet, slightly bitter

Herbal Actions: Anticancer, anticoagulant/antithrombotic, antimicrobial, antioxidant, antiparasitic, hypocholesterolemic, hypolipidemic, hypoglycemic, immunomodulant, nutritive, thyroid tonic

Safety: There is no clinical research on the safety of kelp with pregnancy and lactation, but the long history of traditional use of this seaweed as a food suggests it is safe when consumed in culinary dosages. Use caution with hyperthyroid conditions due to high iodine content. Likewise, large consumption of kelp concurrent with radiation therapy for thyroid cancer may inhibit treatment and is not advised. It is potentially contraindicated with blood clotting disorders and/or anticoagulant medications due to synergistic effects; however, no clinical research has been done on these interactions.

Nutritional Value

Kombu has been a staple in Asian cuisine for centuries, and use of this seaweed has historically been so widespread that it was once even used in Japan to pay taxes. Sugar kelp contains high levels of vitamins A, B, C, and K, as well as the highest levels of potassium of any seaweed. There is also a notable amount of the trace mineral strontium, which is important for the health of bones and teeth. Kombu and sugar kelp have appreciable levels of zinc, calcium, folate, and iodine and are the best seaweed sources of bioavailable magnesium. The iodine in kombu is also significantly more bioavailable than in many other seaweeds. Up to 35 percent of the dry weight is fiber, most of which is sulfated polysaccharides. These species, and other brown seaweeds, contain enzymes that help break down the gas-producing compounds in beans, making them easier to digest.

Adjunct Cancer Therapeutic

Fucoxanthin, the carotenoid pigment of brown seaweeds (like that in sugar kelp and kombu), is behind many of the anticancer actions of this plant. It has been shown in vitro and in vivo to inhibit lymph, prostate, colon, ovarian, breast, and gastric cancer lines' proliferation, migration, and metastasis. It also inhibits angiogenesis (or the formation of new blood vessels in cancerous cell masses) and promotes apoptosis (or death of cancerous cells) by inhibiting nuclear factor kappa-b (NF-κB). NF-κB is a protein complex that controls DNA transcription and cell survival. In the case of cancer, there is unregulated activation of NF-κB, which allows the cells to proliferate without normal apoptosis. Compounds like fucoxanthin that inhibit the activity of this complex are a crucial area of research for cancer prevention and treatment.

Early trials suggest that polysaccharides from *Saccharina japonica* serve as immunomodulators (see Key Terms in this chapter), stimulating both the innate and adaptive immune systems to release nitric oxide and tumor necrosis factor-alpha (pro-inflammatory anticancer molecules) and IL-10, IL-6 (anti-inflammatory cytokines that reduce damage to healthy tissue). There are also indications that pairing astragalus (*Astragalus membranaceus*) with kelp could enhance the anticancer actions of both herbs.

Support for Parasitic Infection

Red seaweeds have traditionally been used for intestinal parasites, and recent scientific research has confirmed that brown seaweed species *Saccharina latissima* and *Laminaria digitata* also contain antiparasitic (also referred to as anthelmintic) compounds. Our current understanding is that this is the result of synergistic effects between eicosapentaenoic acid, alpha-linoleic acid, docosahexaenoic acid, and arachidonic acid, none of which are potent antiparasitics on their own. Like dulse, sugar kelp should only be used as an adjunct therapy for parasite treatment, either alongside herbs such as black walnut and wormwood or taken with prescribed pharmaceuticals. If left untreated, intestinal parasites can wreak havoc on the body's gastrointestinal and immune systems, so best to knock them out as swiftly and efficiently as possible, which often requires allopathic medication.

Antiviral and Antimicrobial

The immunomodulatory actions of *S. japonica* polysaccharide extracts were discussed previously in the context of cancer, but they have been studied in other applications as well. Fucoidan (see Winged Kelp & Wakame for details) also boosts B-cells and helper T-cells, whose roles include secreting antibodies and destroying infected cells. Fucoidan extracts show strong antiviral activity against COVID-19 in vitro, even more so than remdesivir, which was the primary drug for in-hospital treatment of the virus until the prescription drug Paxlovid was developed. Another study comparing the antimicrobial activity of various edible seaweeds revealed moderate activity of kombu against some gram-positive bacteria (which include most of the staph and step varieties and those that cause foodborne illnesses) and yeasts. This all suggests that the next time you get sick would be a great time to add some sugar kelp or kombu into your diet!

Metabolic Syndrome Support

Metabolic syndrome describes a cluster of symptoms including high blood pressure, dysregulated blood sugar, and high cholesterol that increase the risk of diabetes, heart attack, and stroke. All brown seaweeds contain compounds that mitigate the symptoms of (and even prevent) metabolic syndrome. A large body of animal studies has demonstrated the clear antithrombotic/anti-atherosclerotic (anti-clotting), hypolipidemic (blood lipid lowering), hypoglycemic (blood sugar lowering), and hypocholesterolemic (blood cholesterol lowering) effects of consuming daily whole *S. japonica* or its extracts. The small number of human clinical trials concur with these results, thus far. Generally, it takes four to eight weeks of daily intake in order to see significant results.

Systemwide Support

From a broad perspective, the fucoxanthin has systemwide anti-inflammatory action. Cellular metabolism produces normal-occurring unstable molecules, called free radicals, which are kept in check by antioxidants, many of which we procure by eating. In excess, however, free

Seaweed Facial Scrub/Face Mask

MAKES APPROXIMATELY ½ CUP

Exfoliating skin scrubs are so simple but have myriad benefits, including increased circulation, improved lymphatic drainage, clearing of pores, and the stimulation of collagen for smooth and healthy skin. I like to add in other skin-supportive herbs like calendula and roses. Both are astringent (toning) and add a pleasant scent to the scrub. If you would like to make your scrub into a face mask, I recommend adding some ground oats for extra moisturizing.

INGREDIENTS

- 2 tablespoons ground calendula petals
- 1 tablespoon kelp powder (you can buy it ground, or make your own by powdering kelp flakes in a spice grinder)
- 1 tablespoon ground rose petals
- 1 tablespoon ground oats (optional)
- 3 tablespoons bentonite clay

INSTRUCTIONS

1. Combine powdered calendula, seaweed, and roses (and oats, if using) with clay.

2. Mix 1 tablespoon of scrub with some water to form a paste. Gently use to scrub face or other areas of skin. If using a mask, apply evenly to skin and let it sit until clay starts drying, 10 to 15 minutes.

3. Rinse skin, pat dry, and apply additional moisturizer as needed.

HERBAL PARTNER FOR SKINCARE
Calendula

The gorgeous bright orange flowers of *Calendula officinalis* are beloved by gardeners and herbalists alike. Calendula is a flower for the skin, with vulnerary, astringent, and antimicrobial actions that aid with skin issues from cuts and scrapes to rashes and scar recovery. Internally, calendula is a powerful ally for the lymphatic and digestive systems.

Dashi

MAKES 4 CUPS

Dashi refers to Japanese soup stocks made from one or two ingredients, which are then used as the base for numerous other soups. Traditionally, dashi is made by soaking kombu and dried bonito flakes (bonito is a type of fish) in water. I don't love bonito flavor, so I usually omit the fish and add in mushrooms instead. For this, you don't want to heat the kombu. Heating the seaweed starts releasing the gel-forming compounds in the cells and will change the texture, and the taste will shift away from umami toward bitter. Use dashi as the base of miso soup, ramen, pho, and any soup that would benefit from a little more depth of flavor.

INGREDIENTS

1 strip of dried kombu kelp or ½ ounce kombu flakes

6 dried shiitake mushrooms (optional)

4 cups water

INSTRUCTIONS

1. Put seaweed (and mushrooms, if using) in a large pot or other container.

2. Add 4 cups water, plus more if needed to fully cover seaweed (and mushrooms). Let sit at room temperature until water has turned brown, 6 to 12 hours.

3. Strain out seaweed (and mushrooms). Use the dashi right away or keep it in an airtight container in the fridge for up to 1 week.

Seasonal Soup Stock with Seaweed

MAKES 4–5 CUPS

This version of classic European soup stock features vegetable ends, aromatic herbs, mushrooms, and, of course, seaweed. The seaweed is added toward the end of cooking time in order to get the most umami flavor. But if you are pressed for time or need to simplify the steps, it is fine to add the kelp at the beginning (I do this all the time!). This recipe is vegetarian, but you can add chicken or beef bones—be sure to increase the cooking time accordingly. Finally, these ingredient measurements are rough guidelines. Use what you have in the fridge or your garden and alter amounts as you see fit. Steer clear of cruciferous veggies (like broccoli or kale), though—their flavor will overpower everything and become bitter.

INGREDIENTS

- 2 cups chopped alliums (onions and onion skins, shallots, scallions, or leeks)
- 2 large stalks celery, chopped
- 1 large carrot, chopped
- ½ cup mushrooms (whole or leftover stems from another meal)
- ¼ cup fresh burdock root (also known as gobo), chopped
- Several whole cloves garlic (up to a whole bulb)
- Several sprigs fresh thyme
- Handful of fresh parsley
- 3–6 fresh large sage leaves
- 3 bay leaves
- 1 tablespoon whole peppercorns
- 1 teaspoon sea salt, plus more as needed
- 1 small handful of sugar kelp flakes, or 1 large piece of kombu, broken into pieces (about ½ ounce total)

INSTRUCTIONS

If using a slow cooker:

1. Place alliums, celery, carrot, mushrooms, burdock, garlic, thyme, parsley, sage, bay leaves, peppercorns, and salt in the cooker pot and cover with water, leaving about ½ inch headspace at the top. Set to low and cook for 8 to 10 hours, or overnight. Add seaweed in the last 1 to 2 hours of cooking.

2. Once done, strain out the solids. Stock keeps in the fridge for 3 to 4 days, or it can be frozen for several months.

If using the stovetop:

1. Place alliums, celery, carrot, mushrooms, burdock, garlic, thyme, parsley, sage, bay leaves, peppercorns, and salt in a large pot and cover with water, leaving about 1 inch headspace at the top. Bring to a simmer. Turn heat to low and gently simmer for 2 to 4 hours.

2. Turn off heat and add the seaweed. Let sit for 2 to 3 hours.

3. Strain out the solids. Stock keeps in the fridge for 3 to 4 days, or it can be frozen for several months.

HERBAL PARTNERS: FOOD AS MEDICINE

These herbs can support your health in nourishing soups and other healing foods.

Burdock

A rather ubiquitous plant in the northeastern US, burdock (*Actium lappa*) is both loved and disdained due to its enormous taproot. The root is a delicious, slightly sweet bitter, making it an excellent alterative and liver support. It also contains inulin, a prebiotic for our gut flora. Burdock root is known as gobo in Japan and can be found in many dishes, used as any other fibrous root vegetable.

Bay Leaf

From the laurel family, bay leaves have a long history of magic and religious ritual use, as well as being a culinary and medicinal spice. Tea of the pungent bay leaves is used for stomach upset (e.g., gas, bloating, nausea) and clearing phlegm in the respiratory tract. Some people find a poultice can relieve headaches. Traditionally, this herb was also used for treating joint pain and neuralgia, likely due to the warming and circulatory stimulant effects.

Coriander

Coriander, the seed of the cilantro plant, is an often overlooked "kitchen spice rack" medicinal. It is thought to be native to the Mediterranean region but has naturalized across many areas of Asia and Europe. The seeds are citrusy, with just a hint of bitterness. They have been consumed as a carminative (reducing gas and bloating) and digestive antispasmodic across folk medicine traditions. I've featured them in savory dishes with complementary spices, in pickling blends, and as a unique note of citrus in sweeter recipes.

Seaweed Pho

SERVES 4

Pho is a classic Vietnamese soup that consists of a fragrant broth (traditionally made with beef) over noodles and vegetables (or meat) and topped with fresh herbs and bean sprouts. Adding more seaweed to the finished broth further highlights the delicious umami of the dashi base. I love adding wakame because it's tender and slightly sweet, but some tender early season kombu works, too.

INGREDIENTS

- 4 cups dashi
- 4 cups water
- 1 cinnamon stick
- 2 star anise
- 2 tablespoons whole coriander seeds
- 1 tablespoon whole black peppercorns
  1-inch piece of ginger, grated or finely chopped
- 2 lemongrass stalks, bulbs chopped
- 1 teaspoon dried wakame
- 1 large onion, thinly sliced
- 1 carrot, sliced in thin rounds
  Half block of tofu, cut into ½-inch cubes
- 8 baby bok choy, rinsed, bases removed, leaves separated
Rice noodles, soba noodles, or other noodles of your choice
Thai basil leaves, mung bean sprouts, cilantro, lime wedges, for serving

INSTRUCTIONS

1. Bring dashi and 4 cups of water to boil in a medium-sized pot. Add cinnamon stick, star anise, coriander, black peppercorns, ginger, and lemongrass. Reduce heat and simmer for 1 hour.

2. Place wakame in a bowl with cool water to rehydrate as broth cooks. Strain out spices and return broth to the pot. Add onion, carrot, and tofu. Simmer for 15 to 20 minutes.

3. While soup simmers, cook noodles. Set aside.

4. Add bok choy and simmer for 5 more minutes.

5. Turn off heat and stir in wakame. Let sit for 5 minutes before serving.

6. To serve, place noodles in a bowl and top with broth, tofu, and veggies. Garnish with Thai basil, bean sprouts, cilantro, and a squeeze of lime.

Kombucha | Kon-bucha

While we might immediately think of the vinegary, tangy fermented beverage that has rocketed to popularity in recent years, kombucha—or kon-bucha in Japanese—refers to a hot infusion, or tea, of kombu seaweed. (See Bladder Chain Kelp.) To make, simply mix powdered kombu with green tea, and pour hot water over it. Steep for 5 minutes, remove the tea diffuser, and enjoy. You could also do a long steep of this combination, add sugar, and ferment your own kombucha-kombucha!

Seagrove Kelp

O ut in the cold waters of the archipelago of southeastern Alaska is one of the largest seaweed farms in the United States. Seagrove Kelp has been operating since 2018 and grows ribbon kelp (*Alaria*) and sugar kelp (*Saccharina*), also distributing dried kelp online to consumers and wholesalers. The company's founder, Markos Scheer, grew up in the fishing industry in Alaska and spent years providing legal representation to fishermen before entering the seaweed industry, armed with a few kelp cultivation manuals from growers in Maine and an integral knowledge of the Alaskan ecosystem.

A Low-Impact Industry

Kelp farming doesn't require irrigation, fertilizer, or pesticides that displace other species, nor does it necessitate dramatically altering the existing environment, making it a sustainable low-impact industry. And in Alaska there is an enormous amount of potential space available for seaweed farming.

A New Economy

Much of Alaska's economy in the last 150 to 250 years has been based on extracting resources, from gold and furs to oil and natural gas. Markos seeks to position seaweed, and mariculture in general, as a shift in that paradigm that can be used to create value and economically sustain the community.

But the US lacks an established seaweed market. Currently, 97 to 98 percent of cultivated seaweed worldwide is grown on the coasts of Asia and Indonesia, and little of that is sold as a raw product for human consumption. Most macroalgae grown in Korea, for example, is funneled into the very profitable abalone industry as feed for these prized shellfish. And while the vast amount of *Pyropia* grown for sushi nori and SeaSnax–like products is a viable industry, places like Alaska and Maine, which do support kelp industries, lack the infrastructure and market demand to grow it.

Drying Hubs

Markos is interested in how drying hubs can help Seagrove and similar companies break into the larger seaweed market. Seagrove has its own hatchery and underwater farmland, but regional drying hubs would allow them to more efficiently process their own products and support smaller growers to process their crop as well. In his words, it's "Dry or Die!" in the seaweed industry. Drying creates a necessary shift from "producing more [fresh seaweed] than the market can support to not yet producing enough [dried seaweed] for the demand"—which will allow the industry to profitably grow.

Switching to Seaweed

People are not flocking to Alaska to farm seaweed as they are in New England. Alaska's fin fish industry has been in crisis for several years now, but the seaweed industry isn't profitable enough (yet) for people to transition their livelihoods. Markos is excited about the prospect of farming shellfish alongside kelp to build diverse mariculture systems that do not perpetuate the harms of monocropping as seen in land-based agriculture. Biorefining companies are also looking at Alaska. Markos sees biorefining as a strong contender for the future of seaweed farming but cautions that companies need to respect and honor Indigenous sovereignty and land rights, and not tip the balance into extraction and exploitation.

How do we shift the culture toward valuing algae as integral to the identity of people and place?

Many people only buy seaweed as a novelty—it's not a weekly grocery purchase (again, yet!). Markos notes that seaweed is not part of the culture in the United States, even in coastal communities. While lobstering in Maine is understood as a source of identity and pride, for example, Maine is also a hotbed of seaweed farming and harvesting. How do we shift the culture toward valuing algae as integral to the identity of people and place?

With a wave of other companies, Seagrove Kelp is asking these questions and seeking to find ways to remind us (even those of us living far inland) that fundamentally we are all connected to the sea.

Botany & Ecology

Undaria's name comes from the Latin *unda*, meaning "wave" or "ripple." It inhabits the offshore subtidal zones with strong currents. A thick, jointed midrib grows from a dense bundle of ruffled fronds that hide the holdfast. The hollow midrib provides additional buoyancy to the thin, wavy-edged fronds. The fronds usually do not grow longer than 10 feet. While most wakame in commerce is cultivated in China and Korea, it can be found growing wild off the coasts of Japan, southern Australia, Western Europe, and small sections of the north Pacific coast of North America. It is considered an invasive species in many of these places because it is such a robust and adaptable plant. Wakame is one of the most consumed seaweeds in the world and so has some of the most robust clinical research. As such, many of the studies discussed below reference *U. pinnatifida*.

Growth Habits

Alaria also thrives in the low intertidal and subtidal zones of the northern Atlantic and Pacific coasts. It prefers exposed shores with strong wave action. With a single blade growing up to 12 feet, its structure is very similar to *Undaria,* but instead of a ruffled frond at the base, multiple narrow, tapered fronds bearing spores grow from the stipe. *Alaria*'s range is determined by temperature; it can't survive in waters over 61°F (16°C). This species acts as an annual in the intertidal, dying back each fall, and as a perennial in the protected subtidal, with individuals living up to 7 years. *Alaria* should really only be harvested from perennial beds, where there is less pressure on the plants to survive until the next season.

Sustenance for the North Atlantic

Like wakame, winged kelp has a long history with humans and has been eaten as a sea vegetable in Ireland, Scotland, Iceland, and along the coasts of Siberia and continues to be used as animal fodder in northern Atlantic regions. Currently, it is much easier to farm sugar kelp and kombu in the US than winged kelp, so you have to search a little harder for this one!

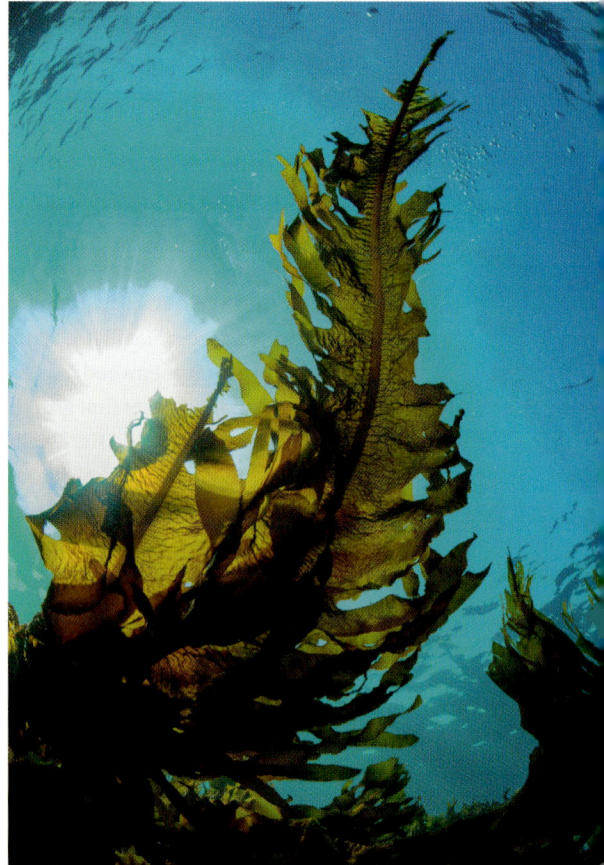

Lush wakame (*Undaria*) grows in the deeper waters of the subtidal zone.

Fucoidan: Potential Anticoagulant

Here is another form of seaweed goo! Fucoidan is a compound in the same class as carrageenan and porphyran found in all brown seaweeds. Formally classified in 1913, fucoidan is a hot water–soluble sulfated polysaccharide with a molecular backbone of fucose (a sugar). Interestingly, fucoidan is also produced by sea cucumbers and sea urchins, though not in the same quantities as marine algae.

Supporting Cell Structure

Biologically, fucoidans add structure to seaweed cell walls and are part of the hydrophilic (water-attracting) coating that helps retain moisture and prevent desiccation at low tide. The sulfate groups on the fucoidan molecules can bind to additional sodium, potassium, magnesium, and calcium, which helps seaweed regulate osmotic pressure within the thallus and adapt to changes in salinity. This means that the cells won't burst or shrivel up when surrounding water salinity changes.

Variation Across Species

The source of fucoidan extracts in clinical research matters because the specific fucoidan structure varies from species to species, and not all fucoidans demonstrate the same actions in clinical research! The timing of seaweed harvest also dramatically impacts fucoidan content, as the concentrations in a specific plant vary throughout the seasons, with the highest concentration occurring at the end of summer and into fall. Depending on species and harvest time, these compounds comprise 5 to 20 percent of the seaweed's dry weight.

Medical Research

Fucoidan has similar structure to certain proteins found in animal tissues (i.e., keratan, chondroitin, and heparin), which has ignited research interest into the parallel use of fucoidans as anticoagulants, antioxidants, antivirals, and immunomodulators. Currently, most heparin (an anticoagulant medication) for clinical use comes from pigs and bovines, but fucoidan extracts from *U. pinnatifida* have demonstrated the same coagulation inhibition and delayed clotting time, suggesting that seaweeds could be a new plant-based source for this lifesaving medication. Fucoidan is also being investigated for antibacterial and anti-allergy effects. Animal research has shown hepatoprotective (liver cell protection) action via inflammation modulation in cases of alcohol-induced liver damage.

MEDICINAL PROPERTIES

Energetics: Cooling, moistening

Taste: Salty, slightly sweet, and generally very mild

Herbal Actions: Anticancer, anticoagulant, anti-inflammatory, antioxidant, antiviral, diuretic (mild), hormonal modulant, hypoglycemic, hypolipidemic, nutritive, thyroid tonic (mild)

Safety: Limited research on pregnancy and lactation, but traditional use suggests they are very safe when used in culinary dosages. Use caution if consuming large quantities concurrent with hyperthyroid conditions. Potentially contraindicated with anticoagulant and anti-arrhythmic medications due to synergistic effects; however, no clinical research has been done on these interactions.

Nutritional Value

In Korea, soup made of wakame is traditionally served postpartum to support milk production and fortify the birthing parent. There is a good reason for this, and for wakame/winged kelp being one of the most popularly consumed seaweeds today. Wakame and winged kelp are high in vitamins A, B2, B3, C, D, and E. These seaweeds are one of the highest food sources of calcium and contain appreciable magnesium, phosphorus, and potassium. Up to 25 percent of the dry weight is proteins. They also contain substantial omega-3s and other essential fatty acids (EFAs). Wakame and winged kelp contain iodine, but in lesser amounts than *Fucus* spp. and *Saccharina/Laminaria* spp.

Adjunct Cancer Therapeutic

Fucoidan extracts are one of the most studied seaweed compounds for anticancer activity. Fucoidan extracts from *U. pinnatifida* and *Fucus* spp. have been shown to halt the progression of metastasis in various cancer cell lines by blocking angiogenesis (new blood vessel growth to feed cancerous cells), restricting cancerous cell migration and adhesion, and inhibiting pro-cancerous cell-signaling pathways. Cancerous cell apoptosis (cell death) has also been demonstrated with extracts from *U. pinnatifida*. As with all seaweeds, the strong antioxidant effects of fucoidan, fucoxanthin, and laminarin in these species boost the anti-cancer effects.

There can also be a synergistic effect between these seaweed extracts and certain cancer treatments, increasing the bioavailability of some drugs and improving the efficacy of others. Clinical trials with whole seaweed supplementation are limited, but the lower incidences of certain cancers in populations who regularly consume these seaweeds anecdotally suggests that they are indeed an excellent dietary addition to mitigate the potential harms of cancer development. Likewise, the long history of feeding wakame to people in states of depletion suggests that consuming these seaweeds alongside or after cancer treatments can help boost vitality and provide extra nourishment that the body needs to heal.

Cardiovascular Tonic

Clearly fucoidan extracts exhibit antithrombotic (anticlotting) and anticoagulant activity in vitro, but several clinical trials have also examined the role of wakame as a functional food for cardiovascular support. A fermented juice made from raw wakame had the effect of lowering blood pressure and impeding platelet aggregation in subjects, suggesting that regular intake could help prevent arteriosclerosis. Another small study using 5-gram capsules of dried wakame daily significantly lowered blood pressure and cholesterol in all study participants after just four weeks. Another trial showed that administration of a strong wakame infusion resulted in suppression of angiotensin-converting enzyme (ACE), which functions to regulate blood pressure by causing blood vessel constriction (see Sugar Kelp & Kombu Kelp for more detail).

In the average cardiovascular system, this is a beneficial regulatory function. However, if a person is experiencing atherosclerosis, high blood pressure, or any other condition characterized by arterial constriction, vasodilation through pharmaceutical or herbal interventions becomes crucial. The bioavailability of fucoidan is still unclear, so combining winged kelp or wakame with other cardiovascular tonics, such as hawthorn, motherwort, and linden, is a good strategy to amplify the cardioprotective effect of seaweed.

Metabolic Syndrome Support

Brown seaweeds, including wakame and winged kelp, offer an array of support for managing (or preventing!) metabolic syndrome. Metabolic syndrome describes a cluster of symptoms including high blood pressure, dysregulated blood sugar, and high cholesterol that increase the risk of diabetes, heart attack, and stroke. As mentioned, fucoidans have cardioprotective and hypotensive actions. Laminarins and alginates are soluble fiber, forming gels that bind with cholesterol and inhibit reuptake in the GI tract. These hydrocolloids also slow the rate of digestion, leading to fewer blood sugar spikes. Fucoidans exhibit additional hypoglycemic effects by slowing intestinal glucose absorption via enzyme inhibition, while simultaneously increasing insulin sensitivity and desired glucose uptake by muscle cells.

Antiviral

Wakame is a traditional remedy for easing coughs and colds, and there is clinical research to back this up. In vitro and in vivo studies show fucoidan extracts from *U. pinnatifida* inhibiting influenza virus and up-regulating immune system function. In another fascinating study, participants were given either fucoidan extracts or a placebo for 4 weeks preceding their annual influenza vaccine. Titers drawn at 5 and 20 weeks later showed consistently higher levels of antibodies in those who took the wakame extracts, suggesting that seaweed supplementation might improve vaccine efficacy. Additional research to tease out the immunomodulatory mechanisms in bioavailable fucoidans from whole seaweed is needed, but eating daily seaweed leading up to vaccine administration seems like a good place to start!

Wakame consumption has also been correlated with improvement of herpes symptoms, and indeed wakame extracts in vitro show inhibition of HSV-1 and HSV-2.

Potential Phytoestrogen Activation

As with *Fucus* spp., there has been some preliminary research done to examine the potential hormone-modulating effects of *Alaria* (and by extension *Undaria*) in the context of estrogen-dependent cancers.

WHAT ARE PHYTOESTROGENS?

Phytoestrogens are compounds in plants whose structure mimics that of estradiol (the primary form of estrogen found in the human body). In a small clinical trial, women took 5 grams *Alaria* daily for 7 weeks and added soy isolates in the seventh week. By the end of treatment, all participants had lowered serum estradiol, increased estrogen metabolism, and enhanced beneficial conversion of soy phytoestrogens into lignans, which are more easily utilized by the body.

POSSIBLE BENEFITS

Generally, phytoestrogens work by binding with estrogen receptors in the body, resulting in weak estrogen-like effects. This sometimes results in a systemic decrease of estrogen production since the brain is receiving signals that all estrogen receptors are being bound and, therefore, the body needs plenty of the hormone circulating. As such, the positive impacts of phytoestrogens (such as those present in *Alaria/Undaria* and *Fucus* extracts) on PMS or menopausal symptoms is still up for debate. We do know, however, that when phytoestrogens (e.g., from soy) can be converted to lignans by the gut flora, they serve as potent anticancer compounds.

In the study referenced above, seaweed significantly increased the conversion of soy phytoestrogens to lignans. The exact mechanisms of the results are still unclear. We do know that seaweed compounds have a high binding affinity for estrogens, reduce cholesterol (a precursor for sex hormone synthesis), and have beneficial impacts on the gut microbiome (necessary for lignan synthesis). With this in mind, it is logical to conclude that winged kelp and wakame supplements may be incredibly helpful in reducing the risk of estrogen-dependent cancers. They may also have dose-dependent applications for managing certain types of PMS or menopausal symptoms.

KEY TERMS
For Understanding Winged Kelp and Wakame Medicinal Properties

Arteriosclerosis: A condition in which the walls of arteries become thick and stiff, compromising blood flow. Atherosclerosis is a type of arteriosclerosis characterized by the buildup of fats, cholesterol, and other substances on the artery walls, causing obstructed blood flow.

Metastasis: The process by which cancer cells spread from their original location to other tissues and organs of the body.

Platelet aggregation: The process by which platelets in blood adhere to each other and form a plug at the site of vascular injury. Excessive aggregation can contribute to the formation of blood clots.

MEDICINAL PREPARATIONS

Like other kelp, these seaweeds are fantastic food-as-medicine. This chapter focuses on vinegars, but I recommend adding them to tea blends as well.

Seaweed and Herb Infused Vinegars

Infused vinegars are one of the easiest and most versatile ways to preserve and enjoy the ephemeral plants and fruits of a season. I like to make batches seasonally and then use them in salads, cooking, and beverages year-round. Seaweeds are perfect here because they are rich in vitamins and minerals, which readily infuse into vinegar. Seaweed also brings a little salt and umami depth to balance the flavor profile. I really like using *Alaria* or wakame here, but you can use kelp or sea lettuce. I've included one recipe for a spring tonic vinegar featuring plants that are bountiful in the spring in the Northeast of the US and one recipe for fire cider. The beautiful thing about infused vinegars is that they are so versatile—please experiment with ingredients and tastes and find out what you love! Just remember that when dried seaweed rehydrates in the vinegar, it will expand quite a bit—so leave space in your jar.

Tincture: Dried plant, 1:5, 30–40%, 1–3 mL per day

Infused Vinegar or Oxymel: Use fresh or dried, whole thallus or flakes.

Infusion: Add flakes to a tea blend.

This seaweed fire cider contains vinegar infused with wakame, citrus, hawthorn berries, Gem marigolds, and ginger.

Spring Tonic Vinegar

MAKES APPROXIMATELY 1 QUART

We make this vinegar in my house year-round using fresh or dried herbs, depending on the season. In spring, though, it is a perfect celebration of warming earth, of waters flowing again, and of delicious greens returning to our kitchen.

INGREDIENTS

3–4 dandelions, roots, leaves, and flowers

½ cup fresh nettles

½ cup additional spring greens (can include cleavers, violet leaf and flowers, sheep sorrel, evergreen tips, plantain, or just more dandelion and nettle)

¼ cup fresh chopped burdock root, or

1 tablespoon dried burdock root

1 large piece of dried kelp or ½ teaspoon dried wakame

Raw apple cider vinegar

INSTRUCTIONS

1. Finely chop dandelions, nettles, and additional greens (leave flowers whole because they look so beautiful). You may want to wear gloves for the stinging nettles.

2. Fill a quart canning jar half to three-quarters full of the herbs. Add seaweed.

3. Pour vinegar into the jar until all herbs are covered.

4. Place a small piece of waxed or parchment paper across the mouth of the jar before screwing on the lid. This prevents the vinegar from rusting the metal.

5. Let the mixture sit for 2 to 4 weeks, shaking as often as you remember.

6. When you're ready to use it, you can either strain out and compost all of the herbs and seaweed, or you can eat them along with your vinegar.

7. Store the mixture in the fridge for up to 6 months.

HERBAL PARTNERS: SPRING HERBS

These spring herbs are perfect additions for seaweed-infused vinegars.

Burdock

Arctium lappa is a gentle, mildly sweet bitter root and an excellent spring herb. Medicinal roots are best dug in spring and fall, and burdock is especially nice in spring when the leaves are still small and manageable and the soil has just thawed. Burdock is an alterative, or what used to be known as a "blood purifier," strengthening the channels of discernment and elimination (e.g., the liver, kidneys, lungs, and lymph). The inulin in burdock is prebiotic and complements the probiotic action of raw apple cider vinegar.

Dandelion

The surest sign of spring in many places are the dandelions (*Taraxicum officinale*) bursting forth with dazzling yellow suns. There are a couple of look-alikes, but if you break the flower stalk and milky sap oozes forth, you've found a dandelion! The whole plant is edible, from root to blossom. The roots are a fantastic bitter to support liver function. The leaves are mineral-rich and diuretic, helping move the water in our bodies. The flowers are tasty and add a pop of color to your vinegar.

Nettle

If you have access to spring nettles (*Urtica dioica*), don't sleep on them! Early spring nettles are tender and often have not developed their full complement of stinging hairs yet. These hairs are actually hollow and contain a mix of compounds including formic acid, the same molecule that makes ant bites itch. Fresh nettles are a classic remedy for seasonal allergies.

Violet

I am partial to including violet (*Viola* spp.) leaf and flowers in all my spring tonics. The ephemeral purple flowers are so beautiful, and the heart-shaped leaves tell you about their deep energetic medicine for the heart and for grief. They are also a gentle and tasty lymphatic and demulcent, adding a soothing quality to formulas. Throw a few leaves and flowers into your vinegar and snack on a few more.

Seaweed Sourdough Bagels

MAKES 8 BAGELS

There is nothing quite like a good bagel. During the COVID-19 pandemic lockdown in 2020, I, like many others, took up sourdough baking. My first missions were English muffins and bagels. Both are easy to make using either fresh sourdough starter or the discard. I think that bagels lend themselves a bit more to savory toppings (though eggs Benedict fans might disagree), so I'm sharing my kelp-infused bagel recipe with you in the hope that you have a delicious savory brunch in your near future.

Sourdough Starter

INGREDIENTS

- ½ cup whole-wheat flour
- 3 cups all-purpose flour
- Water

INSTRUCTIONS

Day 1: In a jar, mix whole-wheat flour with ½ cup water. Loosely attach jar lid or cover with a tea towel and set jar in a warm place for 24 hours.

Day 2: Discard half of the mixture (it may not be bubbling yet), add ½ cup of all-purpose flour and ½ cup water, stir well, and scrape sides down. Cover again and let sit for 24 hours.

Days 3–5: Repeat steps for day 2 each day.

Day 6: Repeat steps for day 2, but every 12 hours instead of every 24.

Day 7: Your starter should be ready! Take a small spoonful and plop it into a glass of water; if it floats, then it has the power to raise dough. If it sinks, discard half and feed again.

Bagels

INGREDIENTS

DOUGH

- 2¾ teaspoons active dry yeast
- 1 cup water
- ½ cup (125 grams) sourdough starter (active or discard, though active will give the best rise)
- 2½ cups warm water
- 1 tablespoon honey
- 2 teaspoons sea salt
- 1 tablespoon powdered dried wakame or *Alaria*
- 4 cups all-purpose flour

WATER BATH

- 8 cups water
- 1 tablespoon sugar
- 2 tablespoons non-diastatic malt powder

Toppings (optional): 1 egg white, lightly beaten; 2 tablespoons sesame seeds; flaky sea salt

INSTRUCTIONS

1. In a large bowl, or the bowl of stand mixer, combine yeast with 1 cup of warm water. Let sit for 10 minutes, until frothy.

2. Add honey, salt, and seaweed. Stir to combine. Add flour and starter and use your hands to bring the dough together (it will be rather stiff). Knead by hand for 8 to 10 minutes, or use a dough hook on a stand mixer and mix on low for 6 to 7 minutes. Cover the bowl with plastic wrap and let it sit at room temperature for 8 to 12 hours.

3. Turn dough out onto the counter and divide it into 8 pieces. Shape each piece into a ball, then place each on your surface and use your thumb to poke a hole straight down the middle of the dough. Pick up and gently stretch the hole to shape a ring. Place the rings on a baking sheet lined with parchment paper. Repeat with remaining dough. Cover the rings with a damp tea towel or loose plastic wrap. Let them rise until nice and puffy, 30 to 60 minutes. They are ready when the dough does not spring back when poked.

4. Preheat the oven to 425°F (220°C). Fill a large pot with 8 cups of water and bring to a boil. Add sugar and malt powder.

5. Boil 3 or 4 bagels at a time (keeping them separate in the water bath) for 2 minutes on each side. (Flip with a pair of tongs.)

6. Remove bagels from the water bath with a mesh strainer and let them rest on parchment paper until cool to the touch. Brush the tops with egg white, dip in sesame seeds, and sprinkle with some flaky salt.

7. Bake bagels for 25 to 30 minutes, or until golden brown. Let cool for 15 to 20 minutes.

8. Store extra bagels in the fridge for up to 1 week, or freeze for up to 3 months.

Nautical Farms

Morgan and Jake, the founders of Nautical Farms, a small-scale kelp farm in Machias, Maine, were not farmers to begin with. They both grew up in Maine—Morgan inland and Jake on the down-east coast, though he moved away to the mountains of Colorado, where they met in college. Jake grew up in a fishing community and knew the waterways, and Morgan had dipped her toes into the kelp product industry through helping to found Akua (a company making kelp-based burgers). With this experience in hand, in 2017 they moved back to Maine and decided to create Nautical Farms together to live according to their values of sustainability and ethical, resilient food systems.

Testing the Waters

In the years since, seaweed farming has taken off dramatically. But in 2017, there were not many folks doing it. Morgan and Jake successfully secured a lease site, but before putting any lines in the water, they spent months building relationships and trust with the fishermen and community. In the commercial fishing industries, folks sometimes raise concerns about seaweed farms, and for good reason. For aquaculture and commercial fishing to successfully coexist, it takes careful planning, communication, and a strong understanding of local waters. Without that, it's easy for lines to be misplaced or for farms to interfere with traditional uses of the waterfront. But Morgan and Jake persevered, and by the time their lines were seeded in the fall of 2017, there was no longer significant community pushback. That spring, they harvested the first round of sugar kelp and have been happily immersed in the cycles of kelp growing ever since.

The Practicalities of Seaweed Farming

I asked Morgan what people might not realize about seaweed farming. She shared the following:

- Seaweed farming does not follow the traditional seasonal cycle—seaweed is actually farmed in winter! Lines are placed in late fall, the plants grow all winter, and they are harvested in spring. As such, seaweed farming in Maine is not exactly a romantic activity. It involves hours on boats in freezing, windy winter weather. You have to be prepared to immerse yourself in the harshest elements.

- The seaweed farming industry remains, much like commercial fishing, a male-dominated field. There are very few farms run solely by women. The first step for women who want to farm seaweed is to gain experience on the water, on boats, or in the fishing industry. These are key skills to set you up for success.

- Seaweed farming may be taking off, but it is not wildly profitable. Like farming on land, there are huge upfront costs, and seaweed farmers often need multiple streams of income or other jobs to make a living.

Sustaining Relationships with Place

Nautical Farms has continued to expand and now sells a variety of food and skincare products, all infused with their kelp. At the time of this writing, they have started to develop fertilizer blends made from scraps that cannot be used for food or cosmetics (e.g., the stipes). Throughout their company's growth, Morgan and Jake have remained committed to the values of sustainable farming and to place that brought them back to the coast.

They dry seaweed in greenhouses and package everything in environmentally friendly containers. They have taken the time to put down roots (and tend the roots from earlier in their lives) and really know the ebb and flow of the town and the sea. They are part of every step of the business, from setting up the lines to harvesting seaweed in all weather and processing it into products for consumers. They're keeping a finger on the pulse of the continually shifting seaweed industry regulations and want to support others in a network of small seaweed farms. Small operations like Nautical Farms are integral to seaweed farming successfully becoming a cornerstone of coastal community economies and social fabric.

The founders, Morgan and Jake, created Nautical Farms together to live according to their values of sustainability and ethical, resilient food systems.

CHAPTER 8

Dulse

Palmaria palmata

At the easternmost point of the United States, in down-east Maine, is the town of Lubec, formerly known as the sardine canning capital of the world, and the 1858 red-and-white lighthouse tower at West Quoddy Head. Here the immense tidal sweep in the bay brings 20-foot tidal changes twice daily. It's like watching the ocean breathe. On the exhale, an entire landscape of plants and animals is revealed, and on the inhale the whole of it is hidden under cold, dark water. The diversity of seaweeds here is incredible, with so many colors and textures all sharing the tidepools, from big kelps down to the smallest epiphytic (a plant growing on another plant) species, as well as species you won't find farther south, including tiny, ribbed fronds of red sea oak (*Phycodrys rubens*), enormous brown shotgun kelp (*Agarum clathrarum*) riddled with holes like Swiss cheese, coarse green tangles of sea emerald (*Chaetomorpha ligustica*), and, of course, the slippery maroon ribbons of dulse (*Palmaria palmata*).

COMMON NAMES

Dulse, dilsk/dillisk, creathnach (Irish), duileasc/duileasg (Gaelic)

Botany & Ecology

The Latin name for dulse, *P. palmata*, contains a double reference to the plant's hand-shaped growth habit (*palma* meaning "hand"). Pacific dulse is a sibling species—*Palmaria mollis* (*mollis* meaning "soft, pliable to the touch"). The mature reddish-purple fronds are about the size of a human hand, 6 to 10 inches long and 4 to 6 inches wide. Dulse (*P. palmata* and *P. mollis*) inhabits the northern Atlantic and Pacific in the mid-intertidal, so you won't see it until the tide is at least halfway out. Dulse is a bit unusual in that it can grow both with holdfast attached to rock and shells and attached epiphytically to the fronds of other larger seaweeds. It's easy to find in its habitat but is not a prolific species like other red seaweeds such as Irish moss. Interestingly, dulse is almost exclusively wild-harvested by hand, as its primary habitat on rocky intertidal boulder fields is not easily accessed by boat or by mechanical harvesting tools. Dulse requires that we use our hands, our palms, to physically touch the fronds, to reach them, in order to harvest, which is part of the medicine of this seaweed.

Feed for Abalone

The oldest record of dulse as a food comes from around 960 CE in Iceland, but it hasn't been a primary food seaweed since the nineteenth century. Traditionally, it was fried in butter or another fat. This seaweed, along with others like *Fucus*, were burned on the shores of Denmark to produce "black salt" (salt crystallized with seaweed ash) for humans and animals alike. The seaweed was also brought inland and spread over the ground to dry and be turned into the soil like any other grassy cover crop. Today, dulse is commercially cultivated on the Pacific coast of North America—but not for humans. It is the primary food for farm-raised abalone that are sold, mostly, to Asian markets. Studies have even shown that the abalone "inherit" some of the medicinal benefits of seaweed by having it as their main food source.

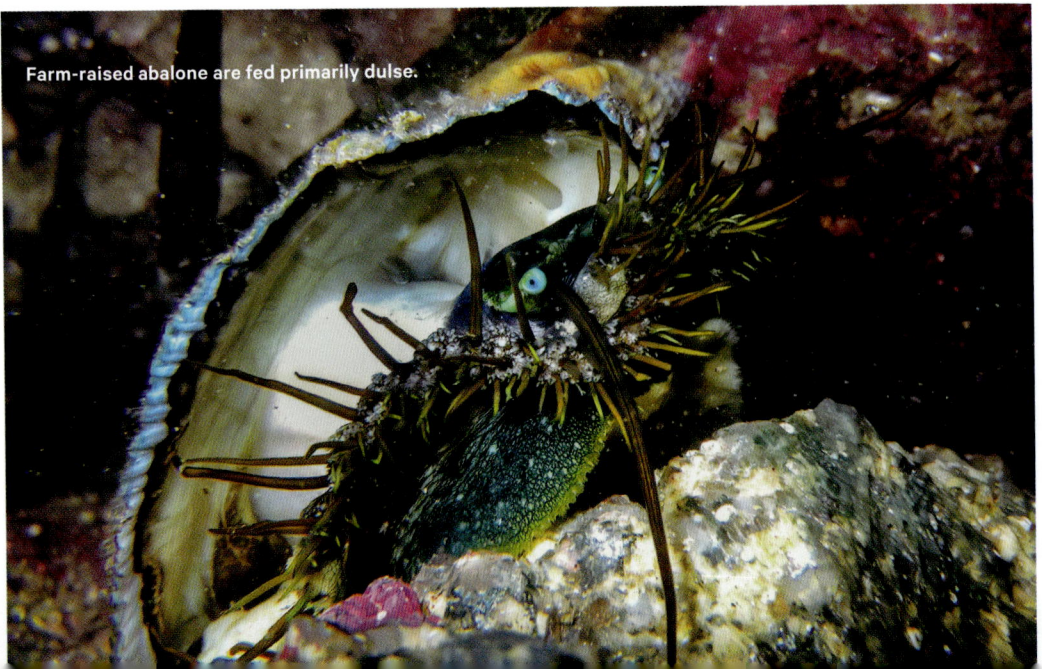

Farm-raised abalone are fed primarily dulse.

wormwood, and oregano, or taken alongside prescribed pharmaceuticals. If left untreated, intestinal parasites can wreak havoc on the body's gastrointestinal and immune systems, so it is best to knock them out as swiftly and efficiently as possible, which often requires allopathic medication.

Metabolic Syndrome Support

Metabolic syndrome describes a cluster of symptoms, including high blood pressure, dysregulated blood sugar, and high cholesterol that increase risk of diabetes, heart attack, and stroke. Cardiovascular protection is important in managing metabolic syndrome, and compounds in dulse can provide some. Like the fucoidans from brown seaweeds, dulse proteins phycoerythrin, phycocyanin, and allocyanin all showed angiotensin-converting enzyme inhibition (though to a lesser extent than brown seaweeds), which results in reduced vasoconstriction and therefore decreased blood pressure. Like nori and laver, dulse also has high levels of the amino acid taurine, which is crucial for cardiovascular and neurological function. The high dietary fiber

content in dulse and other red seaweeds helps lower cholesterol and acts as a prebiotic for beneficial gut bacteria. Clinical trials have had mixed results for *P. palmata* supplementation showing statistically significant reduction in patient blood glucose and cholesterol, so it's best to combine dulse with other brown seaweeds when using a therapeutic for metabolic syndrome or type II diabetes.

Reducing Inflammation for Chronic Conditions

The anti-inflammatory and antioxidant properties of compounds in dulse are rather well studied. Extracts of the phycobiliproteins and chlorophyll a, both from dulse, mitigated acute inflammation and reduced pro-inflammatory cytokine secretion in animal studies. Excess activation of the first-line defense immune system cells, called neutrophils, is a common feature of some chronic inflammatory conditions, such as atherosclerosis. A study with human cells showed that dulse extract disrupted this activation, suggesting that it could be a helpful therapeutic for a variety of neutrophil activity-mediated conditions.

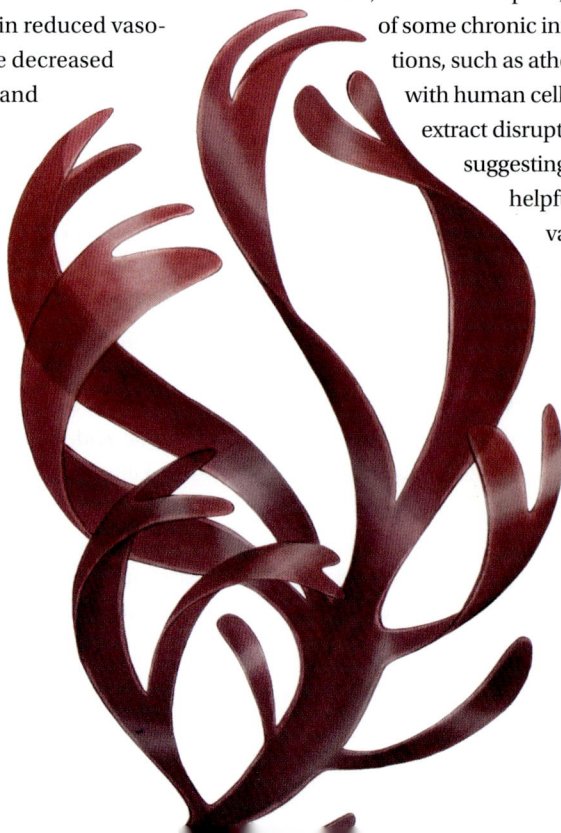

MEDICINAL PREPARATIONS

Dulse is a delicious seaweed and worth experimenting with as a food-as-medicine. This section has a focus on tea blends, but you can use this seaweed in any preparation you like.

Seaweed Tea Blends

Tea always blurs the line between food and medicine. Abby Ferla, owner of Foxtrot Herb Farm, is a seaweed aficionado and herbalist who calls tea "winter salad." When everything dies back in the winter, we turn to our pantries and pull out the dried nettles, raspberry leaf, plantain, milky oats, and other nutritious and flavorful herbs from the last season. A long-steeped infusion of these plants pulls out the vitamins and minerals, delivering a concentrated dose of nutrients in each sip.

Tincture: Dried plant, 1:5, 25–30%, 2–6 mL up to 3 times daily

Capsule: Powdered whole thallus, 0.5–2g per day

Infused Vinegar or Oxymel: Use fresh or dried whole thallus or flakes.

Infusion: Add flakes to a tea blend.

Seaweeds are a perfect addition to these "winter salads" and to tea infusions year-round. Here are some blends that include dulse to get you started, but please experiment with the herbs and flavors that you have access to and that you love best!

A Breath of Salty Air: Respiratory Tonic Tea

Dulse, like other red seaweeds, has a history of traditional use for treating respiratory ailments. This blend combines seaweed with other lung-opening herbs to help you breathe easy. Tip: Use a paper tea bag or strainer so that the tiny hairs on the mullein leaves don't end up in your cup.

1 part dried hyssop
½ part dried spearmint
¼ part dulse flakes
¼ part dried licorice root
¼ part dried mullein leaf
¼ part dried thyme

Use 1 tablespoon per cup of hot water. Steep for 10 minutes. Enjoy 2 to 4 cups per day.

Cardiovascular Tonic Tea

This blend is built around the blood sugar– and cholesterol-regulating effects of dulse and includes herbs with the same set of actions. Try this as a cold infusion (steeped in cold water for several hours to overnight) during the summer.

- 1 part dried hawthorn leaf and flower
- 1 part dried nettle leaf
- 1 part dried tulsi
- ½ part dulse flakes
- ½ part hibiscus flower
- ¼ part cinnamon stick chips

Use 1 tablespoon per cup of hot or cold water. Steep for 10 to 15 minutes. Enjoy 3 to 4 cups per day.

Mermaid Tea

The original version of this recipe comes from Abby Ferla and Ced Clearwater, the herbalists and farmers who run Foxtrot Herb Farm. Abby is a seaweed enthusiast and endorsed these additions of seaweed to one of their signature nutritious herb blends. The shiso and cornflower turn the tea a beautiful dark purple-green that evokes deep-sea gardens.

- 1 part dried peppermint
- 1 part dried shiso (perilla) leaf
- ½ part dried cornflower/bachelor's buttons whole flowers
- ½ part dried nettle leaf
- ¼ part bladderwrack seaweed, flaked or ground

Use 1 tablespoon per cup of hot water. Steep for 5 to 10 minutes. Enjoy up to 3 cups daily.

HERBAL PARTNER FOR THE HEART
Hawthorn

Hawthorn (*Crataegus* spp.) is a primary remedy for the heart and vasculature in both Western herbalism and Traditional Chinese Medicine. The leaf, flower, and berry are all used medicinally. The berries are better for long-decocted syrup, while the leaf and flower shine in tea. Hawthorn helps lower blood pressure, calm the nervous system, and strengthen the physical heart by improving the ability of the cardiac muscle to contract. It's nourishment for the emotional heart as well, an ally for both love and grief.

HERBAL PARTNER IN TCM
Shiso

Also known by the derivative of the Latin name (*Perilla frutescens*), shiso (the Japanese name) is a delicious, slightly spicy, wildly aromatic mint that comes in several varieties. One has leaves with purple tops and green undersides, and another has deep purple leaves with ruffled edges. You might encounter them alongside sushi, or as the purple coloring and flavoring agent in umeboshi plums. Shiso has been used in Traditional Chinese Medicine as an antiparasitic, as well as for digestive and respiratory support. In tea, the flavor becomes smoother and has a clovelike, slightly numbing sensation.

Seaweed tea blends:
Respiratory Tonic (top right),
Cardiovascular Tonic (middle),
Mermaid Tea (bottom left)

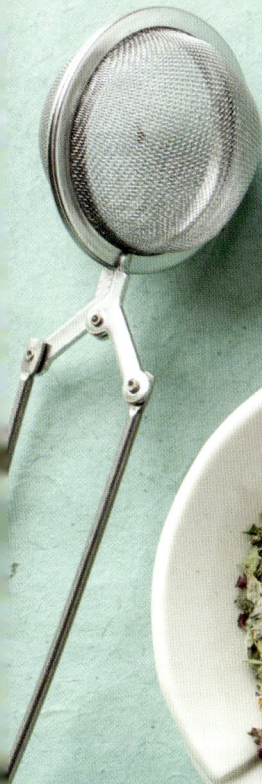

Japanese Curry with Dulse

MAKES ABOUT 3 CUPS

Curries were brought to Japan by British sailors via their colonization of India and adoption of Indian foods. Japanese cooks adapted this recipe over the years, with one of the most famous iterations being katsu curry, which consists of a thick sauce and a breaded pork or chicken cutlet served over rice. I first encountered a recipe for katsu curry from cookbook author Meerha Sodha, who suggests using blended sweet potatoes and carrots as the base for the sauce and breaded eggplant instead of meat. I've adapted the idea to the ingredients that I usually have on hand, including dulse for an extra layer of flavor and a nod to the seaweed culinary traditions in Japan. I enjoy curry sauce over rice with a fried egg and some flash-cooked bok choy, but you can serve it with a protein (chicken, tempeh, pork, tofu) and greens (kale, spinach) of your choice.

INGREDIENTS

- 3 tablespoons olive oil
- 2 medium carrots, diced
- 1 onion, chopped
- 1 large sweet potato, diced
- ½ teaspoon whole coriander seeds
- ½ teaspoon whole cumin seeds
- ½ cup chopped tomatoes
- 4–6 cloves garlic, peeled and pressed or minced
- 1 teaspoon grated fresh ginger
- ¼ teaspoon guntar sannam chile powder (or another chile powder you like)
- 1 tablespoon dulse flakes
- 2 teaspoons turmeric powder
- 1 teaspoon paprika
- ¼ teaspoon ground cinnamon
- 2 cups seaweed soup stock (page 91)
- 2 tablespoons soy sauce
- Salt and freshly ground black pepper

INSTRUCTIONS

1. Heat oil in a large skillet or sauté pan over medium heat. Add carrots, onion, and sweet potato. Cook, stirring frequently, until onion becomes translucent, about 10 minutes.

2. Crush coriander and cumin with mortar and pestle. Add to the pan along with tomatoes, garlic, ginger, and chile powder. Cook for another 5 to 7 minutes.

3. Stir in dulse flakes, turmeric, paprika, and cinnamon.

4. Add stock and soy sauce and bring to a boil. Reduce to simmer and continue to cook until sweet potatoes are soft when poked with a fork.

5. Remove from heat and blend with a standing or immersion blender until smooth and creamy. Taste and add salt and pepper as needed.

6. Serve over rice with the protein and greens of your choice.

Dulse and Olive Tapenade Focaccia

MAKES ONE 9-INCH ROUND LOAF

Two things that I love in this world are breads and spreads. So why not combine them and throw seaweed in the mix, too! Focaccia is a delightfully easy no-knead bread that you can make in a few hours or proof for up to three days. The results are equally delicious, but you'll get a chewier texture from the long rise. Two types of seaweed bring different flavor notes to the tapenade, and the rehydrated laver helps keep the mixture from burning in the oven.

INGREDIENTS

BREAD

- 1 teaspoon active dry yeast
- 1 cup warm water (around 110°F/43°C)
- 2 cups all-purpose flour
- ½ teaspoon kosher salt
- 3 tablespoons olive oil

TAPENADE

- 1 tablespoon dried dulse flakes or fronds
- 1 tablespoon dried laver flakes
- ½ cup pitted olives (kalamata or green)
- 1–2 cloves garlic, thinly sliced
- 3 tablespoons olive oil
- ½ teaspoon lemon juice

INSTRUCTIONS

1. Make the bread: Add yeast to warm water and let sit for 10 minutes for the yeast to bloom.

2. Whisk flour and salt together in a large bowl. Add the water and yeast mixture, and fold together with a rubber spatula until a dough forms.

3. Slick the ball of dough with 1 tablespoon of oil and return it to a clean bowl. Cover tightly with plastic wrap and proof for 1 hour. (Optional: Immediately place the bowl in the fridge and let sit for 12 to 36 hours.)

4. Line an 8- or 9-inch round baking pan with parchment paper and drizzle 1 tablespoon of oil in the bottom.

5. Turn the dough onto a counter and form into a ball. Place in the prepared baking pan and gently press and stretch the dough into a round that is 1 inch away from the edges. Cover with plastic wrap again and let rise for 30 minutes. If removing from the fridge, this second rise will take 2 to 4 hours, depending on kitchen temperature. Dough should puff and spread noticeably. In the last 20 minutes of proofing, preheat the oven to 450°F (230°C).

6. While the bread is proofing, make the tapenade: Place laver in a bowl with cool water for 10 minutes to rehydrate. Drain and chop finely. Finely chop dulse as well if in full fronds.

7. Add dulse, rehydrated laver, olives, garlic, oil, and lemon juice to a food processor and blend into a chunky paste.

8. Drizzle the last tablespoon of olive oil over the dough. Using your fingers, gently stretch the dough to the edges. Spread the tapenade across the top. Use your fingers to "dimple" the dough, pressing it all the way down to the bottom of the pan.

9. Immediately put the dough in the oven. Bake for 25 to 30 minutes, or until top is golden brown.

10. Remove bread from the pan, cool on a wire rack for 10 minutes, and serve.

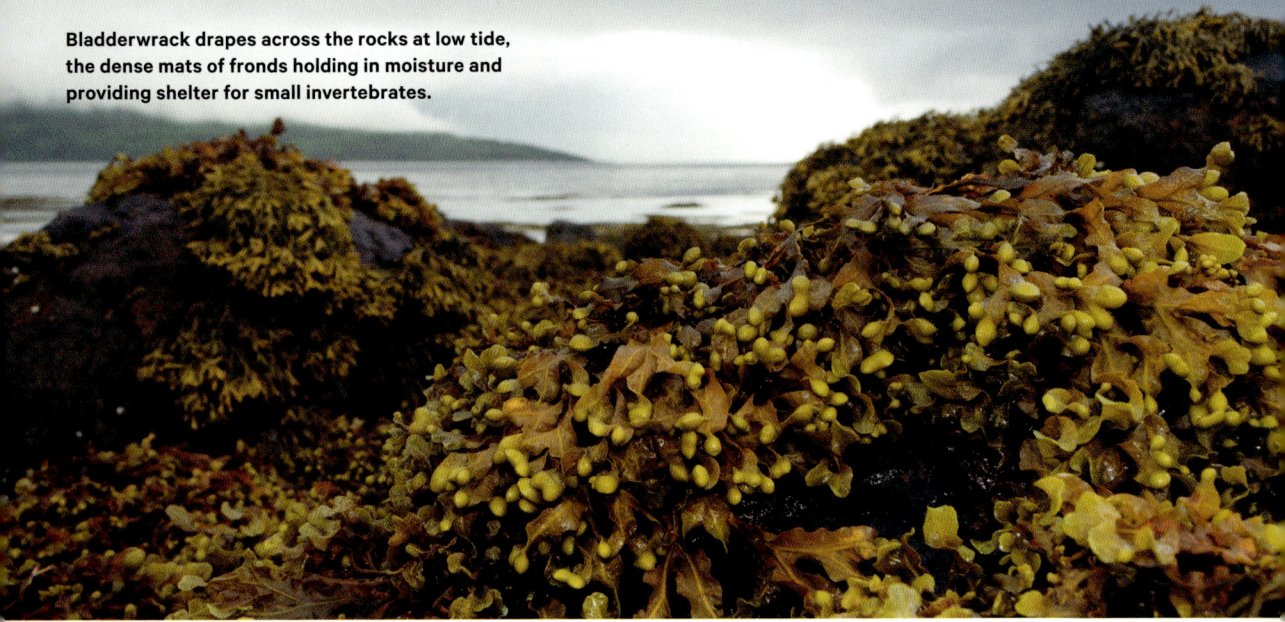

Bladderwrack drapes across the rocks at low tide, the dense mats of fronds holding in moisture and providing shelter for small invertebrates.

Botany & Ecology

Bladderwrack grows in the upper and mid-intertidal zone of the US as far south as North Carolina on the East Coast and from Alaska to the bottom of California on the West Coast. It can also be found on the coasts of Greenland, Iceland, the British Isles, and mainland Europe, as well as the northern coasts of Scandinavia and Russia. Dense mats of the olive brownish fronds snuggle up just above the barnacle bands and drape over the edges of rockweed patches.

Growth Habits

Bladderwrack attaches to rocky shorelines with a holdfast, from which grow forked branching fronds up to nearly 3 feet long. The fronds are distinctive, with (usually) paired air bladders along the midrib of the thallus. The number of bladders is influenced by wave action; more turbulent areas result in fewer air bladders because the seaweed needs less help to be close to the surface of the water, and having fewer bladders reduces friction caused by fast-flowing water. Even with this variability, bladderwrack does tend to prefer

protected shorelines. Mucilage-filled pockets, or vesicles, can be found at the tips of the fronds and house the gametes until later summer, when they are released. The vesicles are textured and covered with pores, or little bumps. When the tide recedes and the vesicles dry and contract, the gametes and mucilage are squeezed out of the pores. When the tide returns, they are washed off into the surrounding water to contact cells from neighboring plants and form new zygotes. Bladderwrack individuals generally live 4 to 5 years.

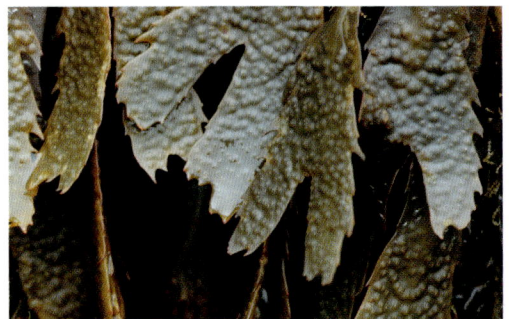

These *Fucus* frond tips display a seasonal variation—the fronds do not always exhibit the bubblelike vesicles.

Versatile and Ubiquitous

Long Island Sound is an extremely populous coast jam-packed with human development. The density with which bladderwrack grows reminds me of the way that people pack into New York City and other urban coastal areas. Bladderwrack knows how to survive and thrive in unpredictable conditions—its fronds are tough and well suited for both racing currents and slow waters. There is a mundanity to bladderwrack—they are the average inhabitant of their environment; they're relatable and reliable. You see them every day on your commute. You might not even notice that you see them because it's such a habitual interaction, until one day, you stop and look closely and introduce yourself. Once you do, you suddenly have a familiar face who's always in your neighborhood, who makes the walk more friendly, whom you now know you'd deeply miss if they were gone.

Iodine

Iodine was first discovered in 1811 by French chemist Barnard Courtois when he was extracting sodium and potassium from seaweed ash for the manufacture of gunpowder. He added some sulfuric acid to the seaweed and observed a purple vapor that condensed into dark crystals. Seaweed was already a documented treatment for goiter, and subsequent scientists quickly observed that giving this seaweed extract to goiter patients produced the same result. In 1896 the presence and crucial role of iodine in the thyroid was confirmed.

Iodine and the Thyroid

The thyroid is a small, butterfly-shaped gland in front of the trachea that is responsible for synthesizing and secreting thyroid hormones, which in turn control a variety of metabolic processes. While there are several chain reactions necessary to produce the active form of thyroid hormones, the initial process of hormone synthesis is fundamentally dependent on the body having sufficient iodine. A classic sign of thyroid dysfunction (often correlated with iodine deficiency) is the swelling of the gland—also known as goiter.

Hypothyroidism and Hyperthyroidism

An underactive thyroid gland (does not produce enough thyroid hormones) often leads to hypothyroidism, which causes a myriad of symptoms, including edema (myxedema in severe cases), fatigue, weakness, decreased blood pressure, constipation, dry skin, brittle hair and nails, depression, cognition issues, and, potentially, weight gain. An overactive thyroid (produces an excess of thyroid hormones), or hyperthyroidism, can lead to rapid weight loss, irregular heartbeat, excess sweating, and irritability. (These symptoms are not an exhaustive list, and thyroid function should be assessed with a full thyroid panel and the support of a healthcare provider.) Some thyroid conditions are autoimmune (such as Hashimoto's thyroiditis or Graves' disease), but not exclusively. When supplementing with iodine, it is important to determine if the condition is underactive or overactive, because iodine increases the activity of the thyroid. As such, it is an appropriate therapeutic for hypothyroidism, but it should be used with caution in hyperthyroidism.

MEDICINAL PROPERTIES

Energetics: Cool, moistening

Taste: Salty, distinctly bitter

Herbal Actions: Alterative, anticancer, anti-inflammatory, antioxidant, antiviral, thyroid-stimulant, diuretic (mild), emollient, hormonal regulation, hypolipidemic/hypoglycemic, immune system regulation, nutritive, prebiotic, radioprotective

Safety: There is no clinical research on the safety of bladderwrack in pregnancy or lactation, but given that this plant has been used as a food for hundreds of years, it is most likely safe in culinary doses. Use caution with hyperthyroid conditions or when used in conjunction with hypothyroid medications. Potentially contraindicated with blood clotting disorders and/or anticoagulant medications due to synergistic effects; however, no clinical research has been done on these interactions. There is one clinical study showing that concurrent administration of amiodarone and *F. vesiculosus* extract compromised the efficacy of the drug: Use caution if taking this and other anti-arrhythmic drugs, as the mechanism of this herb-drug interaction is still unclear.

Nutritional Value

Bladderwrack gets attention for being very iodine-rich, but it also contains solid amounts of potassium, calcium, magnesium, iron, selenium, sulfur, and zinc. Like rockweed, it's particularly high in vitamins C and E. As with all seaweeds, a small serving goes a long way because the nutrients are so concentrated in the plant tissues.

A Mild Alterative

Alteratives are plants that strengthen the channels of elimination and help to gradually bring body systems into balance. These were once commonly known as "blood purifiers" because many of them strengthen the metabolizing and waste-cleaning actions of the liver and kidneys, which all blood passes through as it circulates the body. The polysaccharides in bladderwrack and other brown seaweeds have an affinity for heavy metals and easily bind with them. When eaten, bladderwrack's marine mucilage can bind with toxins in the lower intestine and flush them out of the body. The moistening mucilage along with the high concentration of mineral salts in bladderwrack also make it a mild, soothing diuretic and urinary tonic, which indirectly supports the kidneys, the other primary blood and lymph filtration organ.

Thyroid Stimulant

As discussed, iodine is a critical element for adequate thyroid function. Iodine content in *Fucus* spp. is estimated to be between 0.13 and 0.73 milligrams per gram of dried seaweed, with the highest amount present in *F. vesiculosus*. The daily upper limit for iodine consumption for adult humans is 0.6 mg.

Dosing for thyroid support gets a bit tricky because the iodine content varies in each patch of seaweed and throughout the season as different minerals accumulate in the plant tissues. Additionally, the bioavailability of iodine in the plant varies across species and individuals.

Further, iodine deficiency is a common issue in countries without iodized salt use, so higher iodine supplementation is needed there than in places using iodized salt. With all of this in mind, the following dosages given in the medicinal preparations are conservative. Consult a trained herbalist and/or your healthcare provider if you are considering higher doses.

Topical Uses

The gelatinous insides of the vesicles at the ends of bladderwrack fronds are traditionally used by Indigenous people of North America as a cooling topical remedy for burns, bruises, and sores. You can squeeze some out and apply it the next time you get a sunburn at the beach.

Metabolic Syndrome Support

Brown seaweeds, including bladderwrack, offer an array of support for managing (or preventing!) metabolic syndrome—a cluster of symptoms including high blood pressure, dysregulated blood sugar, and high cholesterol that increase risk of diabetes, heart attack, and stroke. Fucoidans, found in brown seaweeds, have cardioprotective action and antihypertensive effects via inhibition of angiotensin-converting enzyme (see Winged Kelp & Wakame). At the same time, fucoidans from bladderwrack, specifically, have also been shown to exhibit hypoglycemic effects by inhibiting a-glucosidase, one of the enzymes responsible for breaking down starches and complex carbs into simple sugars for intestinal absorption. Lowering this rate of enzyme activity modulates blood sugar.

Another concern in metabolic syndrome and type 2 diabetes is insulin resistance, where muscle, adipose, and liver cells stop responding to insulin, resulting in inadequate uptake and utilization of glucose and dysregulated blood sugar. Animal studies have confirmed that fucoidans from *F. vesiculosus* increase insulin sensitivity and may eventually be a treatment for hyperglycemia in type 2 diabetes.

Cancer Treatment Support

Fucoidan extracts from bladderwrack have been studied for anticancer actions, including inducing apoptosis (cell death) of human colon cancer cell lines, reducing cell proliferation in human breast cancer cell lines, and inhibiting metastasis of human lung cancer cell lines. These extracts also selectively inhibit angiogenesis (new blood vessel formation) in and around cancerous cell masses. Another concern with cancer is the inherent toxicity of treatments, including chemotherapy and radiation. Fucoidans were tested in conjunction with radiation applied to bone marrow cells (BMCs), and the result was increased viability and survival of said cells. The functions of BMCs, which are easily damaged by radiation, include producing red blood cells and white blood cells, the base of the immune system. Additionally, bladderwrack extracts have been linked with increased production of splenocytes (a.k.a. B and T cells for the immune system) from donor BMCs. All told, this humble seaweed may prove to be a source of new (less toxic!) anticancer, radioprotective therapies for people undergoing cancer treatment that may also include stem cell and bone marrow transplants.

Immune Regulation and Antiviral

It bears mentioning that most research on bladderwrack's interaction with the immune system is through the lens of cancer treatment and the examination of the activation of natural killer cells and T cells within the innate and adaptive immune systems, respectively. That said, there have been some recent in vitro and live animal studies examining the dose-dependent effect of fucoidan extracts on inhibiting HSV-1, HSV-2, and HIV infection or replication. Much like with carrageenan in red seaweeds, results were positive. The concentration of fucoidan necessary to prevent viral infection in human tissue is still unclear, but there may be ways to combine carrageenan and fucoidan for potent topical or internal antiviral treatment in the future. Finally, *Fucus* fucoidans seem to increase the efficacy of immune system response to vaccines in the same way that porphyrans from *Pyropia* do, suggesting that we should be eating a good dose of seaweeds before any vaccine administration.

Hormone Regulation

The role of seaweed-derived polysaccharides in hormonal regulation is an exciting new area of research. Cholesterol is necessary for the creation of sex hormones and is a precursor to sex hormone synthesis—and given that brown seaweeds lower cholesterol levels, it stands to reason that consumption of these plants might change sex hormone levels and therefore menstrual cycle patterns. Additional evidence includes that Japanese populations, who consume large amounts of seaweed, have some of the lowest rates of estrogen-dependent cancers and the longest average menstrual cycles ("normal" cycles across all populations range from 23 to 35 days and stay consistent throughout the year). A preliminary pilot study gave three subjects, all of whom had abnormal menstrual cycles, low (700 mg) and high (1,400 mg) doses of powdered bladderwrack in capsules. After 7 months, all subjects reported longer cycles, shorter periods, and reduced endometrial pain. Serum tests confirmed significantly higher progesterone and lower estrogen in all subjects.

Japanese populations, who consume large amounts of seaweed, have some of the lowest rates of estrogen-dependent cancers and the longest average menstrual cycles.

MEDICINAL PREPARATIONS

Bladderwrack is entirely edible, but due to the coarse texture and strong flavor, I rarely use it as a vegetable in culinary dishes. But you can toss whole fronds into any soup or broth, or crush/grind it into small flakes or powder for seasoning. You can also combine it with Epsom salts for a soothing bath soak. In this section, I'm going to focus on using and dosing bladderwrack medicinally for thyroid support, as that is what many people are curious about. The most important thing to note is that, as is the case with any homemade remedies, there is no way to exactly calculate the amount of iodine (or of any of the chemical constituents) present in your medicine without the use of complex lab equipment. The dosages that follow in this section are averages of those that have been passed to me from various herbal teachers and seaweed harvesters.

DIY Capsules

I recommend capsules for bladderwrack because the taste is quite strong and can be off-putting if you're not fully committed to seaweed yet. I often find that it is challenging for new-to-herbalism clients to take tinctures or teas daily, but herbal medicine is only effective if you actually take it, so I offer the capsule method to help overcome that barrier. Making your own capsules is pretty easy; most natural food stores sell empty vegan capsules that you can fill with whatever you wish. You can also find these online. To fill a capsule, open the halves, then scoop each half into your powder and close. You can also use a tiny funnel, folded piece of paper, or small spoon. If you really get into putting together your own capsules, there are devices that hold the capsules and help you

Tincture: Dried plant, 1:5, 25–30%. 2–6 mL up to 3× daily

Capsule: 0.5–2g per day

Infusion: Add flakes to a tea blend.

fill them. The recommended dosage of bladderwrack per day in capsule form is 0.5 to 2 grams. For reference, 1 teaspoon of powder is 3 to 4 grams.

It's easy to fill your own capsules with powdered bladderwrack.

Urinary Tract Tonic Tea

As a mild diuretic, bladderwrack (and other seaweeds) are great tonics for the urinary system. This tea combines soothing and tonifying herbs for a balanced blend. Similarly to the lymph mover blend, if you'd like to cover the herbaceous flavor, add a pinch of mint or tulsi to your cup. If you are formulating for a urinary tract infection, swap the agrimony for one part uva ursi (also known as bearberry) and follow up with a healthcare provider or trained herbalist.

  2 parts couch grass or corn silk
  1 part agrimony leaf
  1 part goldenrod leaf and flower
  1 part plantain leaf
  ½ part bladderwrack, flaked or ground
  ½ part marshmallow leaf

Use 1 tablespoon per cup of hot water. Steep for 10 minutes. Enjoy 2 to 3 cups daily.

Bath Soak

Bladderwrack is so lovely in a bath—especially if you have access to whole fronds of it. They are beautiful as they rehydrate; they float in the water and are chock-full of minerals that help with muscle relaxation and soothing of the skin. Add a handful of fronds to your bath along with some Epsom salts. Elevate the relaxing sensory experience with some lavender buds or rose petals. A slurry of fresh bladderwrack and hot water is also a traditional remedy for joint pain and stiffness. (This could be applied as a poultice, but immersing in a bath is certainly easier and less messy.)

5–10 fronds bladderwrack
¼ cup herbs (rose petals, lavender, or calendula blossoms)

Fill the tub with hot water, adding seaweed and herbs. Let steep for 10 minutes. Herbs can also be placed in a muslin bag and put in the tub to avoid petals clogging the drain.

HERBAL PARTNERS FOR THE URINARY TRACT

These herbs provide support for urinary tract infections.

Agrimony

Agrimonium eupatoria is a classic urinary tract astringent from the rose family. This means it tones tissues that have gotten lax or porous. Agrimony is also anti-inflammatory and antibacterial, making it a great complement to other herbs or medications for chronic pelvic inflammation or infection.

Couch Grass and Corn Silk

These two plants are together because I use them interchangeably as soothing diuretics and urinary tract tonics. We use the rhizome from couch grass (*Elymus repens*), and corn silk (*Zea mays*) is simply the silk from your average ear of corn! Both are best used in a tea, and they taste neutral.

Goldenrod

The Latin name for goldenrod—*Solidago*—means "to make whole." Its usefulness as a urinary antimicrobial is just one of goldenrod's many gifts. Goldenrod is also a respiratory tonic, a remedy for allergies, a gently digestive bitter and carminative, and an aid for sore or injured muscles. If you live in the Northeast of the US, you probably are familiar with the glorious fields of gold flowers that grace us every fall, a last resplendent hurrah before we descend into winter.

Plantain

Humble *Plantago* spp. thrive under our footsteps, filling in gaps in grass, remediating the poor soil of roadsides, and popping up in every crack. Plantain leaf is an incredible remedy because it is both astringent and demulcent, tonifying and soothing tissue in equal measure. It is a primary topical vulnerary (wound-healing) plant that is anti-inflammatory and antipruritic—reducing the pain, itching, and swelling of bug bites and abrasions. Plantain also particularly shines as a remedy for mucous membranes, whether that is in the digestive tract or the genitourinary tract. Slightly salty and mineral-y tasting, the leaves are mild and blend well in tea with almost any other herb, including seaweed.

LAVER
Porphyra

NORI
Pyropia

CHAPTER 10

Nori & Laver

Pyropia spp. & *Porphyra* spp.

A single frond of nori or laver is like a slippery shadow between your fingers, defined but hard to pin down, a solitary thought on a page. An entire basket of them becomes a narrative, rich with the details of each moment harvesting them, of a people tied to the sea. Like the paper we write our stories on, nori and laver are record keepers of coastal peoples. The method for making nori sheets was developed from papermaking techniques in Japan and China. These plants kept our ancestors alive during times of famine.

My ancestors on the British, Welsh, and Norwegian coasts ate laver—the Atlantic equivalent of nori. Traditionally, it was harvested, chopped, and kneaded, then cooked down into a thick, almost black paste called laverbread. This seaweed gets such little interest compared to its cousin, the practically ubiquitous nori that we love wrapped around sushi.

These two seaweeds have strong associations with wealth and richness: Nori was used to pay taxes in Japan, and entire economies in Wales grew out of laver harvesting. Laver and nori carry the wisdom of elders, of memory, of remembering who we are and how we each bring a necessary contribution to the collective experience.

COMMON NAMES

laver (UK), nori (Japan), gim/kim (Korea), luche (Chile), karengo (New Zealand)

Thick mats of green laver can be found growing along the coast.

Ecology & Botany

Nori, laver, and their relatives are some of the seaweeds that have been in taxonomic flux in recent years. Originally all belonging to the *Porphyra* genus, many species of nori have been reclassified into *Pyropia* and will likely continue to shift as genetic sequencing advances. You are likely to find them referenced under either name. Most species of laver remain in *Porphyra*. To reduce confusion, I will use the common names (nori and laver) here. Regardless of where they land in the phylogenetic trees, these are classified as red seaweeds, although they range in color from green to purple to reddish. Both nori and laver grow in the upper intertidal and are incredibly resilient to the stressors of heat and dry air exposure. Many other seaweeds cannot survive the intensity of wave action and sand-scouring that happens here. However, this also means that laver and nori are frequently full of sand and need to be rinsed thoroughly before consuming.

In North America, nori can be found from the Aleutian Islands down to Baja California in Mexico. Laver is represented in North America by seven species that inhabit rocky coastlines from southern Rhode Island to Prince Edward Island in Canada. Most available in commerce is *Porphyra umbilicalis*. Both laver and nori have a tiny discoid holdfast and no stipe. The single frond of each plant is broad, irregularly shaped, and only one to two cells thick (similar to *Ulva* spp.). Both species' full growth ranges from 8 to 12 inches long. Their life cycles are some of the most complex of the red seaweed (as described in the Botany Basics section), which has likely contributed to their tenacity and resilience.

Laver and nori are traditionally hand-harvested seaweeds, as they are small. However, after the breakthrough work of Dr. Kathleen Drew-Baker on the reproductive cycle of nori, it was the first seaweed to be successfully grown at commercial scale. Today, laver species remain a small wild-harvest crop, while nori leads the way in mariculture in Japan, China, and Korea.

At low tide, fronds
of nori shrink and
mold to the rocks
like cling film.

MEDICINAL PROPERTIES

Energetics: Neutral to cool, moistening

Taste: Salty, deeply umami, slightly sweet, and almost nutty when toasted or fried

Herbal Actions: Anticancer/antiproliferative, anti-inflammatory, antioxidant, hypolipidemic, hypotensive (mild), immune regulating, neuro-protective, nutritive

Nutritional Value

Nori and laver have been widely consumed for centuries because of their availability in the inter-tidal as well as their nutritional content. Both contain high levels of protein, magnesium, calcium, iodine, and zinc, as well as A, C, and B vitamins. They also have some of the highest B12 and iron contents of other commonly eaten seaweeds, making them an important diet addition for vegetarians (though there are limited clinical trials to determine the reliability of B12 bioavailability in nori and laver, so vegetarians and vegans should still supplement with additional B12). Due to the high iron content, nori and laver are great additions to the diet for anemic conditions, alongside other iron-rich foods and/or iron supplements. Purple laver is about 40 percent protein by dry weight, which is higher than beef or chicken!

Immune System Regulation and Cancer Support

The immune system touches every part of the body, and immune system dysregulation can cause a variety of ailments. For instance, allergies are the result of a hyperactive immune response to substances that are not actually harmful to the body (from pollen causing hay fever to our own cells attacking normal tissues in the case of autoimmune conditions). With cancer, a hypoactive response causes the body not to recognize certain cell growth as pathogenic

(disease-causing) and so allows cell division and angiogenesis (new blood vessel development to feed these tissues) to go unchecked. Herbs that are immunomodulant, or immune system regulating, can help bring immune system function back into balance.

Porphyran extracts have shown the potential to regulate autoimmune activity, reduce inflammation linked with hyperreactive immune response, and reduce cancer-associated pro-inflammatory cell-signaling activity. Specific anticancer studies have been done with porphyran extracts on human liver, cervical, breast, and gastric cancer cell lines. These studies showed truncation of cancer progression either through dose-dependent cytotoxicity or stimulation of apoptosis (cell death).

Cardiovascular Tonic

Two of the primary issues in the cardiovascular system as we age are high cholesterol and high blood pressure. Laver and nori may help address both. Administration of porphyran extracts has been linked to decreased cholesterol in preliminary studies, suggesting subsequent reduced risk of cardiovascular disease. A high-fiber diet is one of the first steps to reduce cholesterol, and, like all seaweeds, laver and nori are high in fiber. Perhaps equally important, though, are the high levels of the amino acid taurine and omega-3 fatty

Porphyran: A Powerhouse Extract

Porphyran is another complex sulfated carbohydrate (the same class of polysaccharide as carrageenan) that makes up part of the cell walls of *Porphyra* and *Pyropia* species. This compound is responsible for much of the anti-inflammatory, antioxidant, anticancer, immunomodulant, and hypolipidemic effects of *Porphyra/Pyropia* extracts.

The beneficial bioactivity of porphyran is dependent on the molecular weight, which can vary from species to species and over time in an individual. For example, the lower-molecular-weight porphyran extracted from nori that is discolored due to lack of nutrients in the water is not as valuable for consumption (low nutritional value) but has the greatest antioxidant activity. The same paradox often occurs in other "weedy" medicinal plants on land. Plants growing in poor soil or that are stressed produce more of the protective compounds that we use as medicine, making them more desirable even if they look a little ragged. There has been a lot of interest in commercial-scale porphyran extraction, but thus far no technique has produced high enough yields to scale. The low-molecular-weight porphyran derivative oligo-porphyran, however, can be synthesized in a lab and holds promise for future supplements.

Nori and laver have a long history of traditional use as cardiovascular tonics; treatments for high blood sugar, high cholesterol, dissolving cysts, and beriberi (a vitamin B1 deficiency); and therapeutics for bacterial and viral infections. Current research supports these uses, with many linked directly to the phytochemistry of porphyran.

Protein in Seaweeds

Laver is notable for its high protein content, but this is only part of the story. Proteins are large molecules that consist of amino acid building blocks. Our bodies synthesize some of these—the ones known as nonessential amino acids. However, there are nine essential amino acids that our bodies cannot make and must acquire from eating different foods. A food with all nine essential amino acids is called a complete protein. If a food is missing or has very low concentrations of one of these nine amino acids, it is considered an incomplete protein. Seaweeds are all incomplete proteins, containing high concentrations of certain sets of essential amino acids, but not all. As such, seaweeds should be eaten alongside other complete and incomplete protein sources for balanced nutrition.

HERBAL PARTNERS FOR METABOLIC SYNDROME

Cinnamon

Cinnamomum verum, a spice that we find in many baked goods and across cuisines, is also medicinal. Cinnamon is one of the few warming demulcents; the powder forms a thick goo when steeped in a cold infusion. This gelling action traps sugars in the GI tract and inhibits the rapid absorption that causes blood sugar spikes. In this way, cinnamon promotes cardiovascular circulation and has been well established as helping modulate blood sugar.

Artichoke

The leaves of *Cynara scolymus* are the medicinal part of the plant (though of course the delicious taste of the flowers is medicine, too!). Profoundly bitter and cooling, artichoke is a great antihyperlipidemic (aids in lowering cholesterol). The bitter flavor also generally supports digestive function, promoting secretion of bile and liver enzymes.

Tulsi

Also known as holy basil, *Ocimum sanctum* (or *Ocimum africanum*, if you're using the temperate species) is a delightful adaptogen and restorative tonic for various body systems. Tulsi calms the nervous and digestive systems, supports cognition, is antimicrobial and antiviral, and has blood pressure–normalizing effects. It's a fantastic plant on its own or harmonized with others in a formula, and it brings delicious flavor to blends that might be bitter or bland.

Fresh Rolls with Nori

MAKES 8 ROLLS

Spring rolls, also called fresh rolls when not deep-fried, are incredibly versatile and allow you to adjust the ingredients depending on what is in your fridge at a given time. This recipe is a fusion of some of my favorite dumpling and sushi fillings, all wrapped up in rice paper with spicy peanut dipping sauce. There is some prep work involved to get the filling ready, but making the rolls is quick and a great activity for kids, too!

INGREDIENTS

- 1 large sweet potato, baked whole
- 1 sheet nori, crumbled into small flakes
- 8 rice paper wrappers (doubling up the wrapper reduces breakage; if doubling, use 16)
- 1 cup prepared vermicelli or thin glass noodles (or kelp noodles!)
- 1 ripe avocado, thinly sliced
- ½ cucumber, sliced into matchsticks
- 2 cups shredded or thinly sliced iceberg or butter lettuce
- Several sprigs each fresh mint, Thai basil, and cilantro

SAUCE

- ½ cup boiling water
- 1 stalk lemongrass, roughly chopped
- 2 teaspoons toasted sesame oil
- 1 tablespoon grated fresh ginger
- 4–6 cloves garlic, minced or pressed
- 1 teaspoon gochugaru chile flakes (or more if you want more spice)
- 2 tablespoons rice wine vinegar
- 1½ tablespoons soy sauce
- 2 teaspoons tamarind paste
- 6 tablespoons peanut butter

INSTRUCTIONS

1. Remove skin from sweet potato and discard. Smash the flesh into a paste. Add crumbled nori and mix well.

2. Assemble rolls by first dipping a rice paper wrapper in cool water and laying on a flat surface. Wrapper will soften considerably and get very sticky within a minute or so. Smear ½ tablespoon of the sweet potato mixture in a line on the bottom third of the wrapper, leaving a couple of inches on either side. Add remaining toppings in the ratios that suit your taste. Pile it higher than you think you should!

3. Fold up the lower third of the rice paper over the fillings, pressing firmly. Fold in the sides over this initial fold (like folding a burrito), then roll over the remaining rice paper. Repeat for each roll.

4. Make the dipping sauce: In a small bowl, pour boiling water over lemongrass to make an infusion; let sit while cooking the sauce.

5. In a small pot over medium heat, combine oil, ginger, garlic, and chile flakes. Sauté for a few minutes until the mixture starts to brown.

6. Add vinegar and soy sauce, turn heat to low, and cook for 3 to 5 minutes. Add tamarind paste and cook for 1 to 2 minutes.

7. Turn off the heat. Add peanut butter and stir to form a paste. Then slowly add the lemongrass infusion and stir to get the desired consistency. Serve hot or at room temperature, and store any extra in the fridge for up to 1 week.

Laverbread

Laverbread is the traditional Welsh/Northern European preparation of laver. The process takes multiple hours, so do this on a day when you have the time to periodically monitor the stove. Once prepared, laverbread can be stored in the fridge for up to 3 days and frozen for several months. If you have access to fresh laver/nori, you can follow the same process. After rinsing the seaweed well, chop it into small pieces and cook using the method described below.

INGREDIENTS

- 2 cups dried laver
- 4 cups water

INSTRUCTIONS

1. Crush or grind laver (if in large pieces) and place it in a medium-sized saucepan with the water.

2. Bring to a boil, then reduce the heat and simmer for 3 to 4 hours. The mixture should be continuously lightly bubbling. You can use an immersion blender after about an hour to create a smoother texture.

3. Laverbread is ready when enough water has boiled away such that the seaweed clings to itself and pulls away from the sides of pot.

Laverbread

4. Remove from the heat and let cool before using for further recipes or storing.

Lemon Blueberry Laver Oat Cakes

MAKES FOUR 4-INCH CAKES

A classic preparation of laverbread is mixing it with oats to create patties that are fried in some kind of fat. Laver cakes developed a reputation as "famine food" because they were an accessible and cheap way for people in coastal communities to get nutrients when crops failed. With the addition of herbs, spices, and fruit, however, laver cakes become a creative and delicious dish for any meal. For me, these are a deeply ancestral food, and eating them feels like coming home to a part of myself.

INGREDIENTS

1 cup prepared laverbread

⅓ cup blueberries, fresh or frozen and thawed

1 tablespoon lemon juice, plus more as needed

½ teaspoon ground cinnamon

½ teaspoon whole coriander seeds, crushed

¼ teaspoon salt, plus more as needed

½ cup rolled oats (see Note)

1–2 tablespoons butter

Honey, for serving (optional)

Note: Pulse oats several times in food processor for a smoother texture.

INSTRUCTIONS

1. Combine laverbread, blueberries, lemon juice, cinnamon, coriander, and salt.

2. Add oats 1 tablespoon at a time and mix into the laverbread mixture. Taste and add additional salt or lemon juice, if desired.

3. Let the batter sit for 5 to 10 minutes before cooking so the oats can absorb moisture.

4. Heat a nonstick skillet over medium-low heat; when hot, add butter.

5. Form the batter into small patties (they will be a little fragile) and place them in the skillet.

6. Cook the patties on each side until browned, 5 to 7 minutes per side. Serve immediately with honey, if using. The cakes can be stored in the fridge for up to 1 week or frozen for up to 3 months.

HERBAL PARTNER FOR
BLOOD SUGAR REGULATION

Oats

Avena sativa is found everywhere—from skincare products to herbal medicines to everyday breakfasts. In the herbal world, we mostly use the seeds of the oat plant in the unripe milky stage, when they ooze white starchy goo that is a delicious cooling and moistening tonic and restorative for a stressed nervous system. Or we use the plant stalk, referred to as "oat straw," as a deeply nourishing and nutritious component of teas. The fully ripe oats themselves are also medicinal; consuming high-fiber oats helps regulate blood sugar and cholesterol while supporting the gut microbiome and delivering a full range of vitamins and nutrients.

HERBAL PARTNER FOR
LOWER BLOOD PRESSURE

Blueberries

The humble blueberry (*Vaccinium* spp.) is actually a medicinal powerhouse. Blueberries contain high levels of anthocyanins, which are powerful antioxidants. Eating blueberries has been linked with lowered cholesterol and blood pressure, and with increasing insulin sensitivity. Blueberries also support cognitive function and are useful in reducing the onset of neurodegenerative conditions and in healing from traumatic brain injuries and concussions. The leaves of the blueberry's cousin, bilberry, are also widely used as a cardiovascular support. Blueberries and coriander share the same linalool compound, so our brains perceive them as tasting the same, which means that adding some coriander to a blueberry dish amplifies the flavors of both.

Laver Crisp Bread (Knekkebrød)

More akin to an American cracker than bread, these crisps are a traditional recipe from Sweden and are popular throughout Scandinavia. You can swap out seeds for your preferred combination and try different types of seasonings. I am partial to caraway seeds as the primary flavor, making these reminiscent of the dense, dark Scandinavian rye breads. Serve with a sharp cheddar cheese, compound butter, or cream cheese with cucumber and dill.

INGREDIENTS

- 1 cup rolled oats
- ⅛ cup pumpkin seeds
- ⅛ cup sunflower seeds
- ¼ cup ground flaxseed
- ⅛ cup caraway seeds
- ⅛ cup chia seeds
- 1 heaping tablespoon dried laver flakes
- ¼ teaspoon sea salt
- ¾ cup water
- Freshly cracked black pepper

INSTRUCTIONS

1. Preheat oven to 355°F (180°C). Line a 9- by 13-inch baking sheet with parchment paper.

2. Pulse oats in a food processor so they are coarsely ground, but not so fine that they become flour.

3. Pulse pumpkin and sunflower seeds so they are coarsely ground, the same as the oats. Mix oats, all of the seeds, laver, and salt in a bowl with ¾ cup water. Let sit for 10 minutes. Taste and add more salt if desired.

4. Spread seed mixture on a prepared baking sheet, using a rubber spatula to firmly press into a thin layer all the way to the edges. Sprinkle the top liberally with pepper.

5. Score with a knife or pizza wheel.

6. Bake for 30 to 35 minutes, until the edges just start to turn golden.

7. Remove the bread from the baking sheet and let it cool on a wire rack.

8. Store the bread in an airtight container in the fridge for up to 1 week.

Bladder chain kelp, with its characteristic pea pod-shaped pneumatocysts, floats amid other kelps at the edge of the intertidal and subtidal zones.

Botany & Ecology

While firmly in the brown seaweed order of Fucales, the *Cystoseira* genus is in the midst of a taxonomic restructuring. As of this writing, the Pacific species that I originally met as *Cystoseira* (*osmundacea*, *neglecta*, and *setchellii*) have been transferred to the genus *Stephanocystis*. However, when looking in texts, you will likely see the former genus listed. Seaweeds now classified in the *Cystoseira* genus are those that reside solely in the Mediterranean and North Atlantic. For the sake of clarity, I will refer to these seaweeds in this chapter by their common name, bladder chain kelp (even though they are not true kelp, either!).

Growth Habits

The common name is derived from the unique appearance of the upper thallus. The lower thallus arising from the holdfast is perennial and looks like a basal rosette of elongated oak leaves, with deeply lobed fronds that have defined midribs.

The upper thallus is stiff and jointed, with multiple flexible "branches" that bear groupings of air bladders (called pneumatocysts)—they look like a chain of peas in a pod. The reproductive cells develop at the tips of the tiny fleshy, feathery fronds that grow from the ends of the pneumatocysts. At the beginning of winter, when the reproductive season is over, the upper thallus is either shed or torn away by wave action.

The geographical range of this species is relatively narrow, extending from Oregon to Baja California. I have had no trouble finding small individuals in the lower intertidal, but bladder chain kelp doesn't withstand the desiccation of the intertidal well and is more prolific in the subtidal kelp forests. Those growing deeper can reach 30 feet in length. Regardless of location, they need to receive ample sunlight, and they make use of the air bladders to reach as close to the surface as possible.

Umami

In 1908, Professor Kikune Ikaeda at the University of Tokyo isolated granules of the amino acid glutamate from the kombu seaweed used to make dashi broth and coined the term for the singular flavor *umami*. Umami is the fifth basic taste, completing the lineup that includes sweet, salty, bitter, and sour. The word itself is Japanese for "deliciousness." There are three characteristics that define umami:

Spreads across the tongue. Umami taste receptors are found across the tongue, so we experience the taste sensation as one that fills the whole mouth.

Persistence. Umami lingers in the mouth for several minutes longer than the other four flavors.

Promotes salivation. Specifically, salivation lasts for longer sustained time than sour taste.

Why Do We Taste Umami?

The taste experience results from a combination of an amino acid (glutamate, inosinate, or guanylate) and a mineral such as sodium or potassium (e.g., the ever-controversial monosodium glutamate). Just as every taste serves to signal nutrition and safety information to the body, umami tells the body that it has consumed protein. This, in turn, triggers the secretion of saliva and digestive enzymes that facilitate the digestive process. We have umami taste receptors not just on our tongues but in our stomachs, which enhance the cascade of signals that leads to additional digestive enzyme secretion.

Where Do We Find Umami?

Not all sources of umami flavor are protein-rich meats. We find significant amounts of glutamate in tomatoes, asparagus, mushrooms, onions, beets, broccoli, cheese, and, of course, seaweed. The umami flavor found in fish, pork, beef, and poultry, however, comes from the amino acid inosinate. Fermenting and aging food also increases umami. This is partially because intact protein itself has no taste. However, as meat or vegetables age, the proteins in them break down into amino acids, some of which are the glutamate, inosinate, and guanylate, which we can taste. This is part of the magic behind soy sauce, fish sauce, oyster sauce, miso, aged cheese, and cured meats. There is also a science and art to combining the three types of umami to further enhance the flavor. This is why dashi is traditionally made with kombu (glutamate) and bonito flakes (inosinate), and why beef (inosinate) goes so well with mirepoix of carrots, onion, and celery (glutamate). When umami is present, food seems to need less salt, which can be beneficial for dietary concerns and allows other spice and herb flavors to shine more brightly.

Umami flavor tells our bodies that crucial calories have arrived. Adding glutamate-rich seaweed to our food chemically signals to our body that we have taken in nourishment to be digested and absorbed, that we are taking care of ourselves. There is an understandable assumption that seaweed has an overpowering salty or fishy flavor. If we are creative with the application of seaweed to add and enhance umami in dishes, an entire landscape of subtle flavors emerges that tastes nothing like low tide—and is quite literally pure deliciousness.

MEDICINAL PROPERTIES

Energetics: Cool, moistening

Taste: Salty, slightly sweet

Herbal Actions: Anti-inflammatory, antioxidant, antiproliferative/anticancer, nutritive

Safety: There is very limited clinical research on bladder chain kelp specifically, so follow recommendations for other brown seaweeds. Likely safe in pregnancy and lactation in culinary doses.

SEAWEED SCIENCE
Alginates: Good for the Gut . . . and More

Alginates are the most abundant polysaccharides (complex sugar storage molecule) in brown seaweeds, comprising up to 45 percent of dry weight in brown seaweeds. These are water-soluble salts of alginic acids bound to sodium, calcium, or magnesium ions, and they provide structure for the cell walls. Alginates are the compound that gives the seaweed thallus the flexibility to withstand ocean waves and shifting currents. (In contrast, cell walls of land plants contain more rigid compounds, which allow them to maintain upright structure without the support and buoyancy of water.) Alginate is of particular interest as a phycocolloid (plant compound that forms a gel when combined with water) because it can absorb up to 20 times its own mass in water.

Fermented seaweed alginates (administered via capsule) demonstrated prebiotic action both in vitro and in a human clinical trial by increasing production of short-chain fatty acids (SCFA), correlated with increased gut microbial flora. Most importantly, SCFAs are an energy source for gastrointestinal epithelial cells, but they also improve the intestinal mucous membrane barrier for pathogen protection, regulate liver function and insulin secretion, and increase production of necessary hormones. Given the immediate impact on the gut microbiome, alginates may also be a useful supplement to prevent metabolic syndrome as they slow glucose absorption by increasing viscosity of intestinal contents. Additionally, the demonstrated anti-inflammatory and antioxidant effects result in lowered blood pressure, reducing another risk factor for metabolic syndrome.

Most commercial alginates are extracted from giant kelp and rockweed, primarily in the form of sodium alginate. Some of the many applications of sodium alginate include as a thickening and gelling agent for food and beverages; in cosmetics; as a component of antacid medications; for waterproofing and fireproofing materials; and for textile dyes and screen printing. Products with "sodium alginate" often do not list it as such on the label but instead use its food additive code: E401.

Antioxidant for Reducing Inflammation

Reactive oxygen species (ROS), also known as free radicals, are the molecular by-product of many metabolic processes. They easily interact with proteins, DNA, and lipids, causing oxidative damage that can lead to inflammation and other disease processes (e.g., cardiovascular conditions, cancer, metabolic dysregulation). Brown seaweeds contain high levels of phenols and polyphenols, which, in their function as antioxidants, bind with and neutralize ROS.

In a survey of green, red, and brown algae of the Baja California coast, extracts of bladder chain kelp were found to be among those with the highest phenolic content and strongest antioxidant activity, similar to commercially synthesized antioxidants. The same extracts also had considerable nitric oxide (a potent mediator of dysregulated inflammatory response) scavenging behavior, which suggests strong anti-inflammatory potential. In many ways, these compounds act like a cleanup crew in the body, removing excess molecules that would otherwise pile up and cause damage to tissues and organs. These same phenolic compounds have also been shown to exhibit dose-dependent antiproliferative activity on several lines of human cancer cells in vitro. This correlates with the anticancer research being done on extracts from other *Cystoseira* species.

MEDICINAL PREPARATIONS

Bladder chain kelp has many of the benefits of other brown seaweeds and can be used in tea, tincture, or infused vinegar. However, I like to take advantage of this specific seaweed's deliciousness for its beneficial relationship with the gut microbiome. Gut bacteria not only carry out the basic assimilation and absorption activities of digestion while maintaining the health of the gastrointestinal tissue but they also impact the function of our immune system, help regulate metabolism, and relate with our nervous system via the gut–brain axis. Established links exist between gut flora and the delay or progression of neurodegenerative diseases, and between gut bacterial profiles and incidence of metabolic syndrome, type 2 diabetes, and cardiovascular disease. Food introduces novel bacteria to the gut, can act as prebiotics to feed the beneficial species already there, and can provide fiber to keep the GI tract moving functionally.

Tincture: Dried plant, 1:5, 30–40%, 1–3 mL per day

Infused Vinegar or Oxymel: Use fresh or dried whole thallus (fronds and pods are both usable).

Infusion: Add dried flakes to tea blends.

Fermenting Seaweed

Fermentation is one of the oldest food technologies, and aside from being an excellent method of preservation, it allows us to easily access nutrients and to feed our gut microbiome. Every culture has a fermentation practice, from soy sauce, vinegar, and cheese to beer and wine. There is not a strong record of solely seaweed-based ferments throughout coastal regions, but they are a frequent addition to other fermented veggies. You might note that the popular fermented beverage kombucha sounds like *kombu* (see page 93). And you are right, but "kombucha" in Japan is a powdered kombu mixed with green tea and served as a hot infusion. At some point, the term *kombucha* was inaccurately applied to the Western version of acidic fermented tea and sugar, and the name stuck.

Seaweeds are a great base for fermentation, though. They are rich in carbohydrates (bacteria love a complex polysaccharide!) and are full of proteins, vitamins, and minerals that are made more bioavailable through the process. Multiple clinical trials have examined the effects of fermented seaweeds, with positive results on metabolic and neurological functions.

Fermentation is a very accessible practice and can be done in your own kitchen. I have a strong fear of mold and usually leave fermentation projects to other people whose stomachs are less sensitive . . . but kimchi is a ferment that I can and do make. *Kimchi* refers to any number of fermented vegetables, but the most ubiquitous version is made with napa cabbage.

Fermented seaweed

Seaweed Quick Pickles

MAKES 1 PINT

As mentioned previously seaweeds are perfect for pickling, and especially bladder chain kelp with its perfectly crisp and poppable pneumatocysts. From a culinary perspective, pickling is a fantastic way to preserve in-season vegetables, and the salty-vinegar taste brings a joyful burst of flavor to innumerable dishes. Seaweed has high salt content, so you need much less salt in the pickling liquid than a standard pickle recipe. From an herbalist perspective, pickles are also a fantastic way to pack a humble condiment with medicinal benefits. Seaweed can be rehydrated and then pickled, but if using fresh, it is very important to take the time to rinse it and pick out/off any shells, small invertebrates, sand, and epiphytic species before using.

INGREDIENTS

- 1 cup fresh bladder chain kelp (top thallus portion)
- 1 cup water
- ½ cup apple cider vinegar
- 2 tablespoons sugar
- 1 teaspoon sea salt
- 1–2 cloves garlic, thinly sliced
- ¼ cup chopped dill fronds
- 1 tablespoon whole coriander seeds
- 1 teaspoon whole black peppercorns

INSTRUCTIONS

1. Rinse seaweed well and remove any debris.

2. Prepare the pickling liquid by bringing water and vinegar to a boil. Add sugar and salt, and stir to dissolve.

3. Arrange seaweed, garlic, and dill in a pint canning jar. Add coriander and peppercorns.

4. Cover seaweed and spices with the pickling liquid and let it cool for about 30 minutes before putting it in the fridge.

5. The pickles will be ready to eat within 4 hours, but the flavors will develop more and meld together if you let them sit for a couple of days. Use within 1 week.

CHAPTER 12

Rainbow Leaf

Mazzaella californica

~~~

Colloquially called iridescent weed or rainbow leaf, *Mazzaella californica* is distinctive in the tidepool because it shimmers and glows under water unlike anything else. Like the glinting of crows' feathers or the sparkle of fish scales catching sunlight, the wide flat fronds of iridescent weed dance in the waves like darkly holographic sea creatures.

The other incredible gift of *Mazzaella* species is their slime. These seaweeds are particularly high in carrageenan and have a silky-smooth surface. When crushed slightly in water, the slime secretion intensifies and feels delicious on your skin. The carrageenen in all red seaweeds, but particularly in this species, helps to emulsify notoriously tricky homemade creams. Its cooling, soothing, and hydrating properties make it a perfect ingredient in formulations intended to protect and nourish skin.

Note: The medicinal properties section for this seaweed is shorter than others because rainbow leaf isn't frequently used in the culinary world and there is comparatively little clinical research on it. Most studies have been done on this species' Arctic relative, *Iridaea cordata*.

COMMON NAMES

iridescent weed, rainbow leaf; may be listed as genus
Rhodoglossum or *Iridaea* in older texts

Fronds of rainbow leaf shimmer like oil on water in a shallow tide pool.

Botany & Ecology

Mazzaella is a genus of red seaweeds that includes several species exhibiting the trademark iridescence. This shimmering display is not actually linked to any pigment but is structural. When light reflects and refracts through the alternating opaque and translucent cell layers of the seaweed thallus, our eyes perceive these varied wavelengths of light as iridescence. It's much the same as the nacre of abalone (mother-of-pearl). The iridescence may function as UV protection, allowing these seaweeds to live higher in the intertidal zone where solar radiation is stronger.

Wild and Irridescent

Depending on the species, the single lanceolate (tapered) or cordate (heart-shaped) blade may appear reddish purple to yellow-brown and grow up to 3 feet long. (I have typically found much smaller specimens, closer to 1 foot.) The stipe is usually tiny. When in the gametophyte (gamete-producing) generation, the whole thallus will appear rough and bumpy, looking like an entirely different seaweed! In all stages, though, the fronds will be slightly rubbery and have some elasticity when stretched.

The seaweed can be found in protected spots in the low to mid-tidal on rocky coasts from Alaska to Mexico. Rainbow leaf is not a seaweed that is farmed or harvested commercially, so you will have to go looking in the tidepools to find this one. When you do find their glistening fronds, offer your gratitude and thanks before harvesting this gift.

Phycobiliproteins: Antioxidant, Anti-Inflammatory, and More

The green of chlorophyll may be the plant color we are most familiar with, but plants make a number of other "accessory" pigments that serve a variety of purposes other than being directly responsible for photosynthesis. Phycobiliproteins are a type of accessory pigment found exclusively in algae and seaweeds. The two phycobiliproteins that give red seaweeds their characteristic hue are phycocyanin (blue) and phycoerythrin (red). Within algae cells, phycoerythrins absorb light first, then transfer that energy to phycocyanins, which give it to the chlorophylls to create glucose molecules via photosynthesis. Both classes of pigments are used widely as colorants in the food, beverage, and cosmetics industries.

Phycocyanin

Phycocyanin has been studied more extensively, as it is a primary component of the blue-green algae spirulina (*Arthrospira platensis*), which has taken off in the health and wellness industry. Like carrageenan, there are multiple types of phycocyanin: C-phycocyanin occurs in blue-green algae, R-phycocyanin occurs in red algae, and allophycocyanin occurs in both. C-phycocyanin is strongly antioxidant and anti-inflammatory, and it has demonstrated neuroprotective, immunomodulant, and anticancer effects. Concentrated extracts show promise as a potential therapy for diabetes and metabolic syndrome. There also seem to be beneficial impacts on intestinal flora and wound healing. The few R-phycocyanin studies done report anti-allergic and anti-inflammatory actions, specifically in the context of respiratory inflammation.

Phycoerythrin

Phycoerythrin has received a lot of interest in the biomedical field because it is fluorescent, making it useful for detecting tumors in various scanning techniques. Like phycocyanin, there are also variations of this pigment from either cyanobacteria (C-phycoerythrin) or red seaweeds (R-phycoerythrin). R-phycoerythrins are strongly antioxidant and are often included in cosmetics for "anti-wrinkle" and "antiaging" properties. The most common application, though, is in photodynamic therapy for lung, stomach, skin, and oral cancer tumors.

Finally, phycobiliprotein complexes made from both of these pigments are being studied as a therapeutic for cancer, showing promising results in arresting tumor development, stimulating apoptosis (cell death), and improving the safety and efficacy of chemotherapy drugs. This field of research is still just beginning and will no doubt continue to reveal evidence that backs up the long traditional uses of red seaweeds for health and well-being.

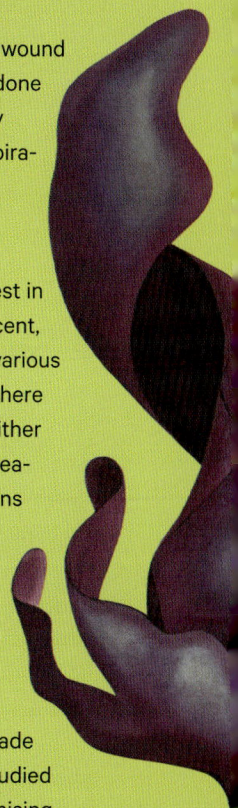

MEDICINAL PROPERTIES

Energetics: Cooling, moistening

Taste: Mild, slightly bitter

Actions: Anti-inflammatory, antioxidant, demulcent, emollient, nutritive, respiratory tonic

Safety: There is no clinical research on the safety of *Mazzaella* in pregnancy and lactation; however, given the traditional use and assessments of other similar red seaweeds, it is probably safe in culinary doses. Use caution if consuming alongside medications, as polysaccharides may slow absorption.

Nutritional Value

There have been limited chemical assessments of *M. californica*, but a study done on a sister species in Japan showed similar B12 levels as nori (*Pyropia* spp.). Like other red seaweeds, rainbow leaf is an excellent source of soluble and insoluble fiber. Like *Pyropia* and *Porphyra*, rainbow leaf seaweed has high levels of the amino acid taurine (four times that of nori!) and polyunsaturated fatty acids (PUFAs.) Taurine is a primary component of bile, which helps to break down and digest fats and fat-soluble vitamins, and it has been shown to potentially reduce hypertension risk through a variety of mechanisms. The PUFA of particular interest is eicosapentaenoic acid, which can only be sourced from oily fish or algae and has been extensively studied for beneficial impacts on heart health.

Respiratory Tonic

Like other red seaweeds, rainbow leaf seaweed contains carrageenan, which has soothing effects on the respiratory system. The gel extracted from cooking it can be taken as a remedy for dry respiratory infections and coughs (see Irish Moss for recipes) or used in cough syrup recipes. The research on phycobiliproteins (page 165) also suggests that this can be a good choice for asthma or allergies that irritate the upper respiratory tract.

Skin Soothing and Healing

Mazzaella is truly a seaweed for the skin. It is just so luxuriously slimy and silky, or what we also call "demulcent." The texture aside, the extracts of this seaweed are highly antioxidant, which supports healthy skin cell growth, repair, and maintenance. *Mazzaella* extracts (or the whole fresh leaf) can be included in your daily skincare routine or used in cases of eczema and psoriasis where there is significant chronic irritation. Research has shown that extracts from Arctic *Iridaea* (a sibling species) are effective against squamous cell carcinoma, which includes some skin cancer lines.

This seaweed is a bit unique in that it's not available commercially, and there is limited research on medicinal applications. As such, I am not providing extract ratios or dosing instructions, and instead I am focusing on its unique value as a topical remedy.

Seaweed Skin Cream

MAKES 15 OUNCES

This is one of my most beloved herbal preparations, adapted from Rosemary Gladstar's classic herbal cream recipe. Creams are notoriously difficult to emulsify, and the carrageenan present in red seaweeds helps dramatically. On that note, you can certainly substitute Irish moss here for the *Mazzaella*, and it will work beautifully. The butterfly pea flowers are here because they contain lovely antioxidants for skincare, but they also turn the cream a beautiful soft shade of blue. Candelilla wax may be subbed for beeswax if you do not use bee products.

INGREDIENTS

- ¼ cup dried *Mazzaella californica* seaweed
- ¼ cup butterfly pea flowers
- 2 cups water
  - Scant ⅓ cup hydrosol of your choice (optional)
- 1 cup infused oil (this can be any oil that you enjoy infused with plants—I like using sunflower oil infused with calendula, violet, and marshmallow. See Herbal Preparations 101 on page 217 for infused oil instructions.)
- 1 ounce beeswax

MATERIALS

- Dry measuring cup
- Liquid measuring cup
- Small saucepan
- Muslin fabric for straining
- Mixing bowl
- Double-boiler saucepan
- Blender
- Silicone spatula
- Container(s) for finished cream

Recipe instructions continued on page 169

What Is a Hydrosol?

Hydrosols, sometimes called plant waters or floral waters, are distillates of aromatic plants and are often created as a by-product of essential oil manufacture. A device called a still is used to produce distillate. Steam is passed through fresh plant material, evaporating and aerosolizing the aromatic volatile oils. This steam is rapidly cooled, then siphoned into a collection vessel. The resulting solution from this first pass of distillation is what we call a hydrosol. If that solution is redistilled multiple times, the volatile oils are concentrated more and more each time until there is pure "essential oil." Hydrosols are generally anti-inflammatory and soothing to tissues. Most are cooling, but some (like ginger) bring warmth. They are a much more sustainable and less extractive way to experience the benefits of aromatic plants than essential oils. Hydrosols can be used topically, consumed internally (a little goes a long way!), or simply used as aromatherapy.

Seaweed Skin Cream, *continued*

INSTRUCTIONS

1. Combine seaweed and butterfly pea flowers in a small saucepan on the stovetop with 2 cups water. Bring to a simmer.

2. Simmer gently for 15 minutes, then remove from the heat and let cool for 15 to 20 minutes. The water should have a nice slimy texture from the seaweed and a gorgeous blue color from the butterfly pea flowers.

3. Strain seaweed and flowers out of the water, using a muslin bag so that you can squeeze as much gooey liquid into the bowl as possible. This is your beautiful carrageenan extraction!

4. At this point, you can either measure out ¾ cup of your carrageenan extraction or combine a scant ⅓ cup of the carrageenan extraction with a hydrosol, if using. Set aside.

5. Place infused oil and beeswax in a double boiler and heat gently until beeswax is fully melted. Set aside to cool for 15 minutes.

6. Pour the carrageenan extraction (plus hydrosol, if using) into the blender and turn on medium-high.

7. *Slowly* pour a thin stream of the oil mixture into the top of the blender. As the amount of oil/wax going into the mixture increases, adjust the blender speed higher as needed. Once all of the oil mixture is added, scrape the sides with a spatula and continue blending on high until the cream is fully emulsified.

8. Pour the cream into containers while it is still warm. Wait to cover the containers until the cream has fully cooled.

9. Store the cream in the fridge for up to 3 months.

Sea lettuce unfurls in the water at high tide.

Botany & Ecology

The *Ulva* genus includes some one hundred species, distributed across the world's oceans. They are in the phylum Chlorophyta, which includes mostly freshwater unicellular algae, and many *Ulva* species thrive in waters close to freshwater influxes. Most species have holdfasts, but a few are free-floating. They are perennial, but we don't actually know how long an individual sea lettuce can live. Generally, *Ulva* grows without a stipe in flat sheets that are just two cells thick, giving rise to their tasty common name—sea lettuce. Seaweed species in which the two layers of cells grow to form a tube that is filled with sea-water often get called gutweed or threadweed. Some of these have recently been reclassified into the genus *Enteromorpha*. All species are relatively small, with the largest only reaching approximately 1 foot long. As their primary pigment is chlorophyll, sea lettuce needs maximum sunlight and generally grows in the shallow intertidal zone, but some species extend into the subtidal. They are delicate and often full of holes and perforations either from wave impact or herbivorous grazing from snails, urchins, and other invertebrates.

There are two species that you might confuse for *Ulva* because they have the same growth habit. *Monostroma* and *Ulvaria,* however, are only one cell thick, rather than two. In order to distinguish, do a fingerprint test—if you can see your fingerprints through the frond while holding it, it's probably *Monostroma* or *Ulvaria*.

Gutweed blankets the rocks and sand at low tide.

With practice, you'll be able to feel the difference between *Ulva* and these other species just by rubbing the blade between your fingers. *Ulva* has had a long relationship with coastal humans around the world, and today it is largely cultivated to feed sea urchins, abalone, and other farmed fish.

Green Tides

Ulva species, like other green seaweeds, have a propensity to proliferate into a bloom when there is excess nitrogen and phosphorus in the water. These "green tides" can smother the coastline and become toxic for marine creatures and humans as they die, decay, and release hydrogen sulfide gas. This is generally more of an issue in locations where the water is warmer and human population is higher, generating greater amounts of nutrient-rich runoff.

Harvest sea lettuce at low tide by removing part of the frond and leaving the holdfast and remaining thallus to continue growing.

Chlorophyll: Seaweeds Have It, Too

We can't talk about algae without giving a nod to chlorophyll. Most of us probably remember from some point in our educational past a biology class with diagrams of little green circular chloroplasts absorbing the rays of the sun and churning out energy for the plant. Chlorophyll is the compound in chloroplasts, and it is the pigment that we perceive as green.

Chlorophyll is lipid-soluble, meaning it's dissolvable in fats, and it consists of a chlorin or porphyrin (not to be confused with porphryan) structure with a magnesium atom at the center. There are at least seven different types of chlorophyll: Chlorophyll a is found in all plants, chlorophyll b in most plants, chlorophylls c1 and c2 are found in algae, and chlorophylls d and f are found in cyanobacteria. *Ulva* and other green seaweeds contain, primarily, chlorophyll b. Brown seaweeds contain mostly chlorophyll c1 and c2, along with the brownish colored fucoxanthin. Red seaweeds contain chlorophyll a, but it is masked by the phycoerythrin and phycocyanin. Studies have shown that chlorophyll from *Pyropia* spp. (nori) is the form most bioavailable to humans, followed by *Ulva* and then *Undaria pinnatifida* (wakame).

Following on the heels of the wheatgrass juice craze, chlorophyll supplements have become wildly popular in recent years, mostly due to their antioxidant properties. However, these supplements are actually made of chlorophyllin, which is synthesized from chlorophyll and has a copper atom at the center rather than magnesium. Clinical research is still inconclusive about the health benefits of chlorophyll supplements, so continue to eat your whole (green) fruits, veggies, and seaweeds!

Ulvans: Difficult to Extract

We've already covered the valuable gel-forming sulfated polysaccharides of red and brown algae (agar, carrageenan, and fucoidan); green seaweeds have a version, too! Ulvans are the compounds found in green algae like *Ulva*, but compared to those in other seaweeds, they have a far more complex structure that has made them a bit challenging to study. In *Ulva* spp., ulvans are the main component of the cell walls, along with three other polysaccharides, and can be up to 36 percent of thallus dry weight.

Ulvans have a wide range of applications across pharmaceutical, food, cosmetic, and agricultural industries, but their use has been limited due to difficulty with extraction and low bioavailability in the human body. A slightly altered version, ulvan-oligosaccharides, are produced in labs and maintain their biochemical activities while being a more stable molecule. The amount and chemical composition of ulvan present in each species of *Ulva* varies by location, season, and harvest time.

Studies have demonstrated that ulvans are anticoagulant, antibacterial, and antiviral. They also can support immune system regulation and help protect the cardiovascular system by reducing total cholesterol and triglycerides, lowering low-density lipids and elevating high-density lipids. They also function as a prebiotic.

Salt

Salt is the flavor of the ocean. It is the flavor of sweat and of tears. It is both a foundation taste and a nutrient that all bodies need to survive. Salty is the first flavor most people notice when tasting seaweed. The white powdery coating that forms on most dried seaweeds is partially salts, and more of it exists in the plant tissue itself. Sodium chloride (NaCl+) is the molecule that tastes salty, but other minerals and chemical compounds can elicit similar flavor experiences. This is why we say that nettles, alfalfa, gotu kola, and other nutrient- and mineral-rich herbs taste salty. You might have experienced salt also having notes of sweet, sour, or bitter. This seems to be due to the way that Na+ ions interact with other taste receptors (i.e., stimulating sodium-dependent glucose transporters) or intracellular reactions between taste buds themselves. To make things even more complex, we have two variations of salt receptors on our tongues, one for "low salt," or the kind of seasoning that makes food taste good; and one for "high salt," or the concentrations that make something inedible or harmful.

An Essential Mineral

Salty is a unique taste because, unlike our experience of sour, sweet, bitter, and umami, with salt we sense on some level that it's a mineral we need for survival. In the body, salt helps regulate blood pressure and is critical in creating and transmitting the electrical impulses that control heart rhythm and muscle movements. When we consume excess salt, the body holds on to water in an attempt to keep the blood from becoming too salty. This increases blood pressure and can put too much strain on arteries over time. (This is the reason why salt has unfortunately been maligned in the medical industry.) On the flip side, not enough salt can lead to a decrease in blood volume, which is why increasing salt intake is a primary therapeutic for postural orthostatic tachycardia syndrome (POTS) and other dysautonomic diseases that present with chronically low blood pressure. Sodium exists in a close relationship with other minerals as well, most notably potassium. Sodium-potassium ion pumps within cell membranes are the driving force behind nerve impulses and muscle contraction.

The salty flavor of seaweed is one of its gifts to the culinary world. You can add pieces of seaweed to a dish instead of table salt, which reduces your overall sodium intake and adds another layer of flavor. Salt makes other flavors pop, which is why a seasoning blend or finishing salt that includes seaweed is absolutely delicious.

Water Follows Salt

From an herbal medicine and energetics perspective, salty flavor (whether that is primarily NaCl or not) usually means that an herb is a diuretic to some degree. In Traditional Chinese Medicine, salt is considered the flavor of the water element and is associated with the kidneys (meridians and channels, not organs). The guiding principle is that water follows salt, and water allows for flow and transformation. Salt therefore dissolves blockages and accumulations, particularly in the lymph, and balances the body's water. It is also a flavor associated with anchoring, calming, and focusing the mind. It's easy to see the medicine of salty seaweed, supporting both flow and stability.

MEDICINAL PROPERTIES

Energetics: Cooling, moistening

Taste: Mild umami, salty, and mineral-y. The dried form is much more bitter than fresh, which is lighter and sometimes almost tangy.

Herbal Actions: Antibacterial, anticoagulant, anti-inflammatory, antioxidant, antiviral, diuretic (mild), emollient, hypolipidemic, immune regulating, nutritive

Safety: There is no clinical data on safety in pregnancy and lactation, but given the long history of traditional use as a food, it is likely safe in culinary doses.

Nutritional Value

Sea lettuce is very similar to land lettuce. It contains concentrated vitamins A, C, and B1, as well as sodium, potassium, and a variety of trace elements. Proteins can account for 15 to 30 percent of dry weight. *Ulva* is notably high in calcium, magnesium, and iron. Magnesium helps the body retain calcium, so foods such as sea lettuce that are high in both are a great supplement for promoting bone health. Sea lettuce has one of the lowest iodine concentrations (and the lowest of any seaweeds profiled in this book), so it can be consumed safely in higher amounts than most red and brown seaweeds. Sea lettuce, with its tender fronds, makes a great raw sea vegetable. Just rinse and add to dishes, or snack on it straight off the rocks! There is a very high amount of dietary fiber in sea lettuce, and it can be prebiotic, feeding the beneficial gut bacteria. One of the only human clinical trials on *Ulva* extracts was designed to test its effects on metabolic markers, and while those results were inconclusive, the patients consuming seaweed extracts had markedly increased concentrations of beneficial gut flora.

Anti-inflammatory and Immune System Regulating

Sea lettuce extracts produce similar anti-inflammatory effects as those observed from red and brown seaweeds, even though the mechanism by which ulvan acts on the body is not as clearly understood as it is with fucoidan (brown), agar (reds), and carrageenan (reds). The human clinical trial mentioned previously also correlated the consumption of ulvan extracts from sea lettuce with decreased proinflammatory cytokines (nonspecific signaling molecules that are part of the inflammation response). Likewise, *Ulva* extracts increase tumor necrosis factor (TNF-alpha), the growth of T and B lymphocytes, and activation of natural killer (NK) cells. Peptides from *Ulva* spp. have been shown to increase production of the anti-inflammatory cytokine interleukin-10 (IL10). IL10 functions to selectively inhibit immune response to pathogens so that healthy tissue is not destroyed; it also regulates the growth of scar tissue. The substances that increase IL10 are excellent allies for healthy wound and tissue healing.

An interesting and notable pharmacokinetic result: When isolated ulvans pass through the digestive tract, they are rendered bioactively inert. The beneficial effects we observe from green seaweeds, therefore, must be due to the synergistic action of ulvans with other compounds and the way those complexes move through the human GI tract. It's a good argument for eating whole seaweed rather than just an extract.

Antimicrobial Inside and Out

Topical Use: Sea lettuce contains a variety of secondary metabolites (molecules formed by a plant's secondary metabolic processes used to adapt to disease, predation, and other stressors), including alkaloids, triterpenoids, steroids, saponins, phenolic compounds, and flavonoids, all of which are collectively antibacterial, anti-inflammatory, and antioxidant. In other more-studied plants, these compounds have clear active roles in accelerating wound healing. In Traditional Chinese Medicine, *Ulva* is used for cuts, burns, insect bites, and wound dressings. In the Americas, Indigenous groups have used *Ulva* to soothe jellyfish stings and to remove boils. The indication to use sea lettuce for topical remedies and infection control is now being explored in the context of wounds that have become infected with multiple-drug-resistant *Staphylococcus aureus* (MRSA). Extracts from *Ulva lactuca* are potent staph bacteria inhibitors in vitro and have shown promising results (comparable to other pharmaceutical treatment) against MRSA. We are still far from hospitals administering seaweed extract poultices to infected wounds, but we'll get there!

Internal Use: In vitro and in vivo animal studies have confirmed that extracts from *Ulva* spp. exhibit antiviral activity against avian influenza, HSV-1, and HSV-2. These same studies show a concurrent proliferation of lymphocytes, suggesting immediate immune system response. Preliminary research has also shown *Ulva* extracts to be effective against COVID-19 by inhibiting viral binding to T cells. It only took about 3 years for carrageenan from Irish moss to go from lab study to FDA-approved nasal spray for COVID-19 prevention, so ulvan extracts from sea lettuce may go a similar route.

> **KEY TERMS**
> ## For Understanding Sea Lettuce Medicinal Properties
>
> **Cytokine:** A variety of cellular messenger substances secreted by immune system cells to relay messages about immune and inflammation response around the body.
>
> **Natural killer (NK) cells:** A special type of white blood cell that can attach to a tumor cell or virus-infected cell and destroy it.
>
> **Peptide:** A molecule made of amino acids that serves as an intermediate building block of whole proteins.
>
> **Pharmacokinetics:** The study of the ways in which drugs or other medicinal compounds move through the body (different from pharmacodynamics, which studies the actual effects of drugs and medicinal compounds).
>
> **Tumor-necrosis factor (TNF):** A crucial cytokine in the immune system that helps promote a healthy inflammation response against pathogens and infection.

Potential Oral Health Supplement: Given the evidence of green seaweeds being anti-inflammatory and antibacterial, it stands to reason that these effects could be applied throughout the body. There has been a singular clinical trial examining the effect of *Enteromorpha* (now *Ulva*) extract on oral health. It was a small sample size, but patients who used seaweed extract mouthwash experienced inhibition and prevention of progression of gingivitis (gum disease) comparable to that of Listerine mouthwash. Perhaps we will soon see mouthwash with ulvans alongside our toothpastes that already contain carrageenan!

MEDICINAL PREPARATIONS

Sea lettuce is a little more versatile than your average head of iceberg or romaine. Its mild flavor and delicate texture make it easy to incorporate into tea blends, but it also shines in decocted preparations like the iron syrup on the facing page.

Tincture: 1:5, 30–40%, 1–3 mL per day

Infused Vinegar or Oxymel: Fresh or dried, flakes or whole thallus. The mild flavor combines well with many other herbs.

Tea Blends: Add ½ teaspoon flakes to your favorite tea blend.

HERBAL PARTNER FOR IRON DEFICIENCY
Yellow Dock

Yellow dock (*Rumex crispus*) is one of those plants that grows everywhere and anywhere, popping up in sidewalk cracks, on top of septic drain fields, and in poor soil along the road. It is indeed yellow, and the leaves often develop rust-colored spots that make it easy to identify. The root does contain some iron, but yellow dock primarily functions as a plant that helps our bodies take up iron more efficiently. This is why yellow dock and iron-rich molasses are the foundation of any iron-boosting syrup.

Iron Syrup with Sea Lettuce

MAKES 3 CUPS

Iron syrups are a classic herbal preparation made with iron-rich molasses paired with herbs that enhance the body's ability to absorb iron. This is a great tonic for mild iron deficiency or an additional support alongside iron supplements for clinical anemia. It's also a lovely boost if temporary anemia or "blood deficiency" occurs during menstrual bleeding. This recipe takes advantage of the high levels of iron in *Ulva* spp. for a deeply nutritive and fortifying tonic. (Please consult a healthcare provider to routinely test for iron levels if chronic anemia is a concern, as there are many potential underlying causes.)

INGREDIENTS

- 1 ounce dried yellow dock root
- ¾ ounce dried hawthorn berry
- ½ ounce dried nettle leaf
- ⅓ ounce dried dandelion leaf
- ⅓ ounce dried dandelion root
- ⅓ ounce dried *Ulva* seaweed flakes or whole leaf
- 5 cups water
- ½ cup blackstrap molasses
- ½ cup honey (you may choose to just use another ½ cup of molasses rather than honey)

Spices for additional flavor boost, such as 1 tablespoon fresh ginger, 2 cinnamon sticks, or 5–10 cardamom pods (optional)

INSTRUCTIONS

1. Combine yellow dock, hawthorn berry, nettle leaf, dandelion leaf and root, and seaweed in a medium-sized pot with 5 cups of water, and bring to a simmer.

2. Simmer until the water volume has reduced by half, about 1 hour.

3. Remove from the heat and squeeze through a strainer lined with muslin/cheesecloth. You should have about 2 cups of liquid. (The roots and berries absorb quite a bit of liquid that won't get pressed out, which is why you don't end up with exactly half of 5 cups.)

4. Add molasses and honey, and stir to dissolve.

5. Decant the syrup into a clean jar and label. Store in the fridge for up to 1 month. Take 1 tablespoon daily for a general boost in iron intake, or 1 tablespoon daily during your menstrual cycle if using for menstruation-induced anemia and fatigue.

Midsummer Fresh Herbs and Seaweed Infused Salt

MAKES APPROXIMATELY 1 CUP

This recipe was inspired by a delicious herbal finishing salt that I used to buy when I lived in Boston, and by a finishing salt that I had once on a dish at Lil Deb's Oasis in Hudson, New York. As with many of the herbal recipes in this book, feel free to play with flavor combinations and find the ratio of ingredients that suits your palate best.

INGREDIENTS

- ½ ounce fresh basil, chopped
- ⅓ ounce fresh sage leaves, chopped
- ⅓ ounce fresh Tangerine or Lemon Gem marigolds (or substitute the zest of 1 lemon)
- 2 cloves garlic, chopped
- 1 cup kosher salt
- ¼ cup ground sea lettuce and nori, combined
- 1 tablespoon dried rose petals, ground
- 1 teaspoon dried lavender

INSTRUCTIONS

1. Preheat the oven to 200°F (95°C) and line a baking sheet with parchment paper.

2. Put basil, sage, marigolds, garlic, and salt in a food processor. Pulse until the herb pieces are only slightly larger than the salt granules and components are well combined.

3. Spread the herb and salt mixture in an even layer on a prepared baking sheet and bake for approximately 20 minutes, until herbs are fully dried.

4. Combine the herb and salt mixture with seaweeds, rose petals, and lavender. Taste a pinch and see if you'd like to add more of anything else. Store the infusion in an airtight container for up to 1 year. Use it in savory dishes, or sprinkle it on everything and anything.

Macadamia Nut "Hollandaise Sauce" with Sea Lettuce over Roasted Vegetables

MAKES 2 CUPS

Several years ago, I had a vegan hollandaise sauce at the now-closed Fox and Fig Cafe in Savannah, Georgia. Upon returning home, I went on a mission to re-create that magical sauce. I learned to cook many macadamia nut–based sauces when we lived in Hawaii, so that was the obvious choice, but cashews will work just as well. Adding powdered sea lettuce brings the salt and amps up the umami flavor, and it feels like just the right pairing with the macadamia nuts grown in Hawaii. If you want things saltier, add some black sea salt—the sulfurous taste contributes to the egg flavor. You will need a high-powered blender to get the creamy texture right for this sauce.

INGREDIENTS

SAUCE

- 1 cup raw macadamia or cashew nuts
- 1–2 cups almond or oat milk
- ¼ cup lemon juice
- ¼ cup nutritional yeast
- 1 tablespoon ground sea lettuce
- 1 tablespoon white wine vinegar
- ½ tablespoon turmeric powder
- 1 teaspoon garlic powder
- ¼ teaspoon freshly ground mustard seeds
- ¼ teaspoon (or more) guntar sannam chile powder
- ¼ teaspoon ground white pepper

ROASTED VEGETABLES

- 1 pound asparagus
- 2 red bell peppers, sliced into strips
- ½ pound new potatoes or red potatoes, cubed
- Olive oil
- Salt and freshly ground black pepper
- Paprika

INSTRUCTIONS

1. Preheat the oven to 400°F (200°C). Place nuts in a small bowl and pour boiling water over them. Let them soak for 1 to 6 hours.

2. While nuts are steeping, roast vegetables: Separately toss the asparagus, peppers, and potatoes with oil to coat. Sprinkle all with salt and black pepper. Place asparagus on one baking sheet and peppers and potatoes on another. Roast asparagus for 20 minutes and peppers/potatoes for 40 minutes, or until edges are crisping and potatoes are cooked through.

3. Drain nuts and transfer them to a high-powered blender. Add almond milk, lemon juice, nutritional yeast, seaweed, vinegar, turmeric, garlic powder, mustard seeds, chile powder, and white pepper, and blend on high speed until silky smooth.

4. Transfer the sauce to a pot and bring it to a simmer. Cook, stirring frequently, until sauce thickens, 15 to 20 minutes.

5. Serve hot over roasted vegetables. Add a sprinkle of paprika.

6. Store extra sauce in an airtight container in the fridge for up to 1 week. This also freezes well.

CHAPTER 14

Caribbean Sea Moss

Gracilaria spp.

Sea moss, Jamaican sea moss, and Caribbean sea moss *all* refer to a variety of *Gracilaria* species that grow and are widely cultivated in warm Caribbean waters. But it's important to be aware that the seaweed health products industry erroneously uses "Irish moss" and "sea moss" interchangeably, even though these are two morphologically different red seaweeds! It is infuriating. The health benefits of *Chondrus crispus* (Irish moss) and *Gracilaria* spp. (Caribbean sea moss) are very similar, but it does a disservice to muddle the two. As an herbalist, I may use peppermint and lemon balm in much the same way for much the same reason, but they are still different plants!

Because of my frustration with the misleading labeling, I've spent more time getting to know *Gracilaria* than I otherwise would have. Off the coast of Massachusetts, the cold-water species can be found in the lower intertidal or washed up in wrack lines. When you pick up sea moss, there is a sturdiness and density to the branches of the thallus that is unlike most of the other seaweeds described in this book. It is almost animal-like, reminiscent of the bodies of sponges or tunicates (sea squirts). Extracts from *Gracilaria* spp. and their close cousins are present in much of our food, cosmetics, and pharmaceuticals. Soils are inoculated with "seaweed sap" from *Gracilaria* to promote the growth of crops. This seaweed plant is so much a part of our animal bodies, and our culture, and most of the time we don't even know it.

COMMON NAMES

Caribbean sea moss, Jamaican sea moss

Botany & Ecology

Gracilaria is another seaweed genus that is currently in taxonomic flux. Most members are morphologically similar, the sporophyte and gametophyte reproductive phases look similar, and hybridization is common, so genetic sequencing will no doubt continue shuffling the genus and species designations in years to come.

Gracilaria are generally warm-water seaweeds, but species grow up and down the Atlantic and Pacific coasts, with *Gracilaria edulis* being the most heavily cultivated tropical species. They all have relatively small disc-shaped holdfasts that either attach to rocky surfaces or are stabilized in sandy or muddy substrate. Since *Gracilaria* does not always attach to hard ground, it grows best in sheltered waters, estuaries, and bays. Interestingly, it is a seaweed that often prefers brackish waters to true salt water. The thallus is made up of a tangle of thick and fleshy branches that are usually glossy smooth and can be up to a foot long. While dark reddish purple in the ocean, most sea moss in commerce has been dried and bleached to pale golden tan. *Gracilaria* has adapted to thrive in a wide range of temperatures and salinities and can be found in the intertidal and subtidal. In mariculture, it's grown on long singular lines, tube nets, and floating mats. While many products bear the label "Caribbean sea moss" or "Jamaican sea moss," most *Gracilaria* is cultivated on the coasts of Southeast Asia and Chile. The majority of sea moss that actually comes from the Caribbean is wild-harvested rather than farmed.

Agar: What's in a Petri Dish

Agar is a sulfated polysaccharide (just like carrageenan, ulvan, and fucoidan) found in the cell walls of red seaweeds. The molecule is actually a combination of the polysaccharide agarose, which contributes the gelling action, and agaropectin, which increases the gel strength. Agar is hot-water soluble, solidifying at temperatures lower than 104°F (40°C), and so is stable at room temperature. Agar is also thermoreversible, meaning it can be remelted and cooled back into a stable solid without damaging the properties or texture. When combined with sugars, potassium, or calcium, the gel becomes even sturdier, which is useful in the manufacture of foods and pharmaceuticals.

Uses

Agar (along with carrageenan) is one of the primary emulsifiers and stabilizers used across the food, pharmaceutical, and cosmetics industries. Agar has been in use since the sixteenth century as a commercial hydrocolloid (gelling agent) and was the first of these plant-derived compounds to be approved by the FDA. Perhaps the most famous use of agar is in lab petri dishes for growing out bacteria strains. In the late nineteenth century, German scientist Robert Koch was trying to find a sterilizable, high temperature–stable medium on which to grow bacteria and provide evidence for germ theory. Potato slices and broth gelatin weren't cutting it. Finally, Angela Hesse, the wife of one of the lab team members, used her knowledge of cooking with agar to develop a perfect bacteria-growing medium. With that, Koch was able to grow out the bacteria that cause tuberculosis and cholera—and pave the way for the future of microbiology.

Gelidium spp. were originally the primary source of medical-grade agar, but due to unsustainable farming and harvesting practices, a general shortage has both driven up prices and increased demand to find alternative sources. *Gracilaria* spp. are now also being tapped for extracts and account for over half of agar production worldwide.

MEDICINAL PROPERTIES

Energetics: Cooling, moistening

Taste: Slightly salty, bland

Herbal Actions: Antibacterial, anti-inflammatory, antioxidant, antiproliferative/anticancer, demulcent, emollient, hypoglycemic, laxative, nutritive, respiratory tonic

Safety: No clinical research yet exists on the safety of *Gracilaria* in pregnancy or lactation, but given that this plant has been used as a food for hundreds of years, it is most likely safe when consumed in culinary doses. Use caution when consuming the gel close to taking medications, as mucilage may inhibit absorption. Frequently consuming large amounts of sea moss may cause lower GI upset in sensitive individuals.

Nutritional Value

Caribbean sea moss is very palatable, especially when well rinsed and blended into gel form. The agar and carrageenan contribute significant insoluble fiber. Caribbean sea moss contains high levels of vitamins A, B2, C, and E, and minerals, including calcium, iodine, sodium, magnesium, zinc, potassium, and amino acids. Sea moss also contains omega-3 fatty acids, which are compounds necessary for brain and body function but that are not produced by the human body. Seaweed is not an adequate source by itself for all of the omega-3s the body needs, but it is a great daily boost in addition to other sources.

The recommended serving size to start is 1 to 2 tablespoons daily of sea moss gel, but you can eat up to ½ cup, depending on the concentration of seaweed in the gel and your GI tract sensitivity.

Respiratory Support

In general, demulcent herbs like seaweeds are considered soothing for the respiratory tract because of shared nerve channels between the nerves in the GI tract and those in the lungs. When ingested, even though the plant constituents do not actually touch lung tissue, they have a "reflexive" soothing and tonifying effect on lung mucus, cilia, and smooth muscle. *Gracilaria* spp. have a long history of use in South America and the Caribbean for coughs, colds, bronchitis, and asthma. The traditional preparation is quite flavorful: Sea moss is cooked with vanilla, cinnamon, and nutmeg, with mango, banana, or soursop added for additional flavor and nutrients. Just like its northern cousin Irish moss, Caribbean sea moss relaxes the respiratory muscles and helps reduce inflammation of the airways, making it a helpful remedy for acute respiratory conditions and a tonic for chronic asthma, COPD, and even long-term smoking or smoke exposure.

GI Support

Internally, the agar in sea moss gel is wonderfully soothing to an irritated GI tract, whether that irritation presents as ulcers, heartburn, sore throat, or ulcerative colitis, or as symptoms of IBS. When taken with plenty of water, the gelatinous carrageenan and agar also become a gentle bulking laxative that can help support bowel movements in periods of constipation. This is one of the oldest documented uses of *Gracilaria* in the Caribbean and mainland North and South America. In current research, extracts of an agar-derivative called agarose-oligosaccharide show promise as a prebiotic, feeding favorable gut bacteria.

The Aphrodisiac Question

One of the most controversial claims made about sea moss is its power as an aphrodisiac. This is likely because of its traditional use in Caribbean medicine as a remedy for erectile dysfunction and its use in Asian medicine for testicular swelling and penile concerns. Plants and herbs that have been labeled as aphrodisiacs or sexual tonics generally have one or more of the following effects: (1) improving circulation, both in the core and out to extremities, including genitals; (2) relaxing the nervous system so that you can be present in the moment and experience pleasure; and (3) hormone modulation to help with arousal.

Caribbean sea moss does not have any of these effects directly, but the high levels of vitamins and nutrients help all body systems function more effectively, from energy levels supported by adequate iodine intake for the thyroid to skin health supported by B vitamins. The demulcent and mild diuretic actions also support urinary tract health, and in bodies where the urethra runs through the penis, a healthy urinary tract is important to pain-free arousal. The zinc content may potentially help support sex hormones that are key to libido, but there is no research evidence as of yet to back any of these claims. That said, just as you can make lube with carrageenan from Irish moss, the same can be done with *Gracilaria*, which can certainly add some ease and enjoyment to sex!

MEDICINAL PREPARATIONS

Like Irish moss, this seaweed can be tinctured with low alcohol, but it is a much better food-as-medicine plant. You can use it interchangeably in any of the medicinal recipes given for Irish moss.

Tincture: Dried plant, 1:5, 30–40%, 1–3 mL per day

Capsule: Powdered whole thallus, 0.5–2 g per day

Infused Vinegar or Oxymel: Use fresh or dried, whole thallus or flakes.

Infusion: Add flakes to a tea blend.

Sea Moss Gel

MAKES APPROXIMATELY 2½ CUPS

While you can certainly buy premade gel, it's super easy to make your own. You just need a high-powered blender, such as a Vitamix or something similar. Dried sea moss is relatively easy to find in markets if you live in south Florida; otherwise, you can order online from a variety of companies. The key to preparing pleasant-tasting sea moss gel is to soak, drain, and rinse the seaweed multiple times before blending. Sea moss gel can be blended into the recipes in this chapter or eaten on its own as a daily supplement for whole body support. Recommended dose: 1 to 2 tablespoons daily.

INGREDIENTS

2 ounces dried sea moss
2–3 cups water, plus additional for soaking

INSTRUCTIONS

1. Rinse sea moss and pick out any shells and other extraneous seaweed bits (depending on the harvesting practices, you might end up with small pieces of other seaweeds that don't get cleaned off the *Gracilaria* fronds). If there are significant salt deposits on the seaweed, soak it for 5 minutes, then drain and rinse twice before proceeding to the next step.

2. Place sea moss in a large bowl and cover it with cold water. Chill in the fridge for at least 3 hours but preferably overnight, so that it rehydrates and expands.

3. Transfer sea moss to a high-powered blender and add 2 cups of cold water.

4. Blend on medium to high speed, adding more water as needed to achieve the consistency of a smooth gel. If blended too little, your gel will be rather chunky and less easy to incorporate into other food or beverages. There is no danger of overblending, though!

5. Store the gel in an airtight container in the fridge for up to 1 week, or freeze it in cubes for later use.

Sea Moss Smoothies

I am a self-proclaimed smoothie connoisseur. My dad prides himself on knowing all the best mango trees in the county to harvest from and keeps a chest freezer stocked with mango year-round. My parents raised me and my siblings on smoothies made from these fresh mangoes, bananas, and orange juice. I've made thousands of smoothies over the years. These are the components for what I consider a perfect smoothie:

- A creamy base (bananas, avocados, or sea moss gel)
- A balanced palate of sweet and tangy fruit and veg (e.g., blueberries, raspberries, pineapple, mangoes, acai, pitaya [dragonfruit], green apples, cucumbers, spinach)
- A little zip or fresh note (e.g., mint, basil, fresh ginger, kombucha)
- Neutral alt-milk (e.g., oat, almond, coconut, macadamia, soy, rice)

The most important thing to remember is that a smoothie is only as good as its raw ingredients. If you have an overripe avocado and a green mango, it's going to taste chalky and cloying. Get to know your produce and you'll have a delicious, blended meal every time.

Green Smoothie

SERVES 2

We eat with our eyes first, and this greenish teal smoothie delights me before I even take a sip, every time.

INGREDIENTS

2 cups oat or almond milk

½ cup passion fruit pulp, fresh or frozen

1 banana, frozen and cut into pieces (freeze when ripe, but not to the point of brown spots)

3 inches cucumber, peeled

⅓ cup ripe avocado

¼ cup frozen mango

¼ cup frozen pineapple

Small handful baby spinach

5–10 fresh mint leaves

2 tablespoons sea moss gel

1 teaspoon blue spirulina powder (optional; to make the green color really pop!)

¼ teaspoon grated fresh ginger

INSTRUCTIONS

1. Place all ingredients in a high-powered blender. Mix on medium to high speed until thick and creamy. Add more liquid if needed.

2. Serve immediately.

What About Spirulina?

You might have noticed that spirulina does not have a feature in this book. Although it is called blue-green algae, it is not technically a seaweed. It's actually a form of cyanobacteria. The three species *Arthrospira platensis*, *A. fusiformis*, and *A. maxima* were formerly classified in the genus *Spirulina*, and the name has stuck. Blue-green algae has historically been harvested as a food by peoples across Central America and Africa, and it continues to be an economically and nutritionally critical resource in Chad, Ethiopia, and Kenya. It is as vitamin- and nutrient-dense as any of the macroalgae, and there is extensive clinical research on the health benefits of consuming spirulina. You can find spirulina in both the green form (whole cyanobacteria) and the blue form (pure phycocyanin extracted from the organism). The green form contains a more complete nutrient profile and a slight earthy flavor, while the blue form is the single compound and tastes much milder.

Pink Smoothie

SERVES 2

Pitaya (dragonfruit) has become widely available in the frozen fruit section of grocery stores, and I love making a bright pink smoothie with it to brighten the day.

INGREDIENTS

- 2 cups almond or oat milk
- ½ cup frozen pitaya/dragonfruit
- 1 banana, frozen and cut into pieces
- ⅓ cup frozen raspberries
- ⅓ cup passion fruit pulp, fresh or frozen
- 2 tablespoons sea moss gel
- ¼ teaspoon grated fresh ginger

INSTRUCTIONS

1. Place all ingredients in a high-powered blender and mix on medium to high speed until thick and creamy. Add more liquid if needed.

2. Serve immediately.

Passion Fruit Haupia with Mango

SERVES 2-4

Haupia is a coconut milk–based pudding dessert found throughout Polynesia. The pudding relies on starch (arrowroot is traditional) or gelatin to thicken and set. One of my favorite setting agents for this is agar (in place of gelatin), which keeps the recipe vegan and infuses a little seaweed goodness. Agar does set a little firmer than gelatin, so the result has less "bounce." Tart and sharp fruit flavors balance the rich creaminess of the coconut milk. You can use fresh or frozen fruit.

INGREDIENTS

- ⅔ cup coconut milk
- ⅔ cup passion fruit purée, frozen juice cubes, or fresh passion fruit pulp with seeds removed
- 1–2 teaspoons coconut sugar or honey
  Pinch of salt
- 1 teaspoon agar powder or 2.5 grams agar flakes
- 1 fresh mango, sliced or cubed, or ½ cup frozen mango thawed, for serving

INSTRUCTIONS

1. Heat coconut milk and passion fruit in a small saucepan until just boiling.

2. Add sugar to taste and salt; stir to dissolve.

3. Add agar, stir to dissolve, and continue to simmer for 2 minutes.

4. Pour the mixture into an 8- by 4-inch glass baking dish (or similar container) and let sit for 30 minutes to cool and begin to set.

5. Chill for 2 hours or overnight before serving. Cut into cubes and top with fresh mango to serve.

Rest and Digest Gummies

MAKES 24 GUMMIES

This blend was developed when I was thinking about how I could translate the calming and relaxing feeling of Celestial Seasonings' Sleepytime tea into a gummy that would be a quick bite of calm. I work with a lot of clients on sleep support, and I'm always trying to get creative about ways to ingest herbs that aren't a full mug of tea right before bed or an alcohol-based tincture. This recipe uses dried herbs; you can, alternatively, use fresh skullcap, wood betony, chamomile, and maypop (passionflower).

Have fun with the molds for these next two recipes—craft stores and cooking supply stores sell them, and there are endless options online. I use ones that hold about 1 teaspoon of liquid per gummy.

INGREDIENTS

- 4 tablespoons dried ashwagandha root
- 2½ cups water
- 2 tablespoons dried maypop (passionflower)
- 2 tablespoons dried skullcap
- 2 tablespoons dried wood betony
- 3 tablespoons dried chamomile
- ¼–½ cup honey, adjust according to your preferred sweetness
- 1¼ teaspoons agar powder or 3 grams agar flakes
- Cornstarch, for dusting

INSTRUCTIONS

1. Place ashwagandha in a medium-sized pot with water. Bring to a boil. Reduce to a simmer and decoct for 30 minutes.

2. Add maypop, skullcap, and wood betony. Simmer for another 15 to 20 minutes.

3. Remove pot from heat, add chamomile, and cover with a lid. Let sit for 15 to 20 minutes. (This infuses the chamomile without turning it bitter—most other herbs in this blend are bitter.)

4. Strain out the herbs and return the liquid to the pot.

5. Add honey to taste and stir to dissolve while heating gently to a simmer. Taste and adjust sweetener accordingly.

6. Add agar powder and stir to dissolve; simmer for an additional 2 minutes.

7. Remove the pot from the heat and let the mixture cool for 5 minutes. Carefully pour it into the molds of your choice. Fill to the top. (A spoon or pipette works best for this.) Alternatively, you can pour the mixture into an 8- by 10-inch glass dish, let it cool, then cut it into 24 even portions.

8. Chill the mixture in the fridge for 2 hours. Remove the gummies from the molds and dust them with cornstarch to prevent them from sticking to each other.

9. Store in an airtight container in the fridge for up to 1 month.

Elderberry Gummies

MAKES 48 GUMMIES

Elderberry syrup is a staple in herbalists' pantries, offering an amazing immune boost and with the added benefit of tasting delicious. I love making elderberry syrup throughout the year. Sometimes, though, syrup gets boring. Enter the herbal gummy! Gummies are perennially popular, and they rely on some kind of gelatin or gelatin-like substance. For vegan gummies, I like to use seaweed-derived agar. It sets firmer than gelatin, so the gummies are a little less "gummy" than the standard variety.

This recipe is for an elderberry syrup-turned-gummy, but you can swap ingredients and experiment with creating your own blends.

INGREDIENTS

- 2 inches fresh ginger, finely chopped
- ⅓ cup dried elderberries
- 2 tablespoons dried echinacea leaf and flower
- 1 tablespoon dried rose hips
- 2½ cups water
- ½ cup honey
- 1¼ teaspoons agar powder or 3 grams agar flakes
- Cornstarch, for dusting

INSTRUCTIONS

1. Place elderberries, echinacea, rose hips, and ginger in medium-sized pot. Add 2½ cups water and bring to a boil. Reduce to a simmer and cook until the liquid has reduced by half, 30 to 60 minutes. You should have about 1 cup of liquid, since the berries and rose hips absorb quite a bit.

2. Strain out the herbs and return the liquid to the pot.

3. Add honey and stir to dissolve, while heating gently to a simmer.

4. Add agar powder and stir to dissolve; simmer for an additional 2 minutes.

5. Remove the pot from the heat and let it cool for 5 minutes before carefully pouring the liquid into molds of your choice. (A spoon or pipette works best for this.) Use 1 teaspoon per mold.

6. Chill the mixture in the fridge for 2 hours. Remove the gummies from the molds and dust them with cornstarch to prevent them from sticking to each other.

7. Store the gummies in an airtight container in the fridge for up to 1 month. Take 1 to 2 daily for preventive health and 3 to 5 daily if sick. Because this uses safe food-as-medicine ingredients, you don't need to worry too much about precise amounts per serving.

HERBAL PARTNERS FOR THE IMMUNE SYSTEM

Add these herbs to your recipes when you want to give your immune system a boost.

Rose Hips

The fruit of the rose plant (*Rosa* spp.), the hips are packed with vitamin C and other antioxidant compounds. These are a great addition to any immune system–boosting blend and are also lovely for helping with body tissue repair after inflammation from illness or injury. Purchase them dried or harvest your own at the end of the growing season—and remember to leave some for the birds and other creatures that use these as winter food.

Echinacea

Easily the most famous antimicrobial and immune-stimulating herb, *Echinacea* (*purpurea* or *angustifolia*) is a necessary addition to formulas for illness. It is an excellent lymph mover, alterative, and antifungal. The characteristic zingy, mouth-numbing, and saliva-stimulating sensation of echinacea is due to alkylamides, which are also found in spilanthes flowers. This compound is most concentrated in the roots but is also present in the flower heads and, to a lesser extent, the leaves. You can use the root in syrups, but it is extremely bitter, so I recommend the leaf and flower here. Fresh is best, but dried works, too.

HERBAL PARTNERS FOR REST AND RELAXATION

Use the following herbs to aid with sleep and soothe the nervous system.

Maypop (Passionflower)

We call *Passiflora incarnata* (and other species) maypop because they burst into bloom in May in the southeastern US, where they are endemic. The Spanish and Portuguese names (*maracuya* and *maracuja*) both come from the Indigenous language name *mara kuya,* or "the fruit that serves itself." *Passionflower* was the name given to these gorgeous vining plants by Christian mercenaries to Central and South America, replacing an entire history of a plant with religious settler-colonizer imagery. While multiple recipes in this book call for the sweet and delectably tart fruit pulp, the entire vine is medicinal. The leaves, vines, tendrils, and flowers all help to calm the nerves. Maypop is considered a sedative nervine, though I find that in small doses it is perfectly appropriate for daytime use to calm a frazzled nervous system. It has a bitter flavor and a sort of earthy, fruity, slightly musty smell.

Ashwagandha

Withania somnifera is an Ayurvedic herb well known for being an adaptogen and supporting sleep. Ashwagandha is not a sedative herb, though; it works much more effectively as a long-term balancer of circadian rhythm, supporting quality sleep and nourishing your energy levels during waking hours. It's an excellent anti-inflammatory and immune system–regulating herb. The beneficial compounds in ashwagandha are even better absorbed by the body when consumed with a fat, so you could try making these gummies with a little coconut or oat milk (similar to the haupia recipe on page 194).

Wood Betony

Not to be confused with *Pedicularis* spp., which also has the common name of *wood betony*, *Stachys officinalis* is a fantastic nervine that connects to our gut–brain axis. As a bitter plant, it's cooling, with downward-moving energy that seems to have an affinity for tension held in the solar plexus region. It's also an herb specifically indicated for dissociative states, trauma recovery, and PTSD/C-PTSD. Wood betony brings us back into our bodies so that we can, quite literally, rest and digest.

Skullcap

Of all the nervines, *Scutellaria lateriflora* might be the one I reach for most. It's gentle yet profound; it calms racing thoughts and twitching muscles, bringing a sense of ease to the body and the mind. Safe for kids and elders and alongside medications, it blends easily with other nervines and sleep-supportive herbs. I highly recommend growing this perennial friend; the gorgeous purple flowers are a delight each season, and the aerial parts can be made into tea, tincture, or oxymel.

Freshly harvested hijiki. Hijiki has many uses in Traditional Chinese Medicine and is delicious, though it should be eaten in moderation.

Hijiki *Sargassum fusiforme*

Phaeophyceae
COMMON NAMES: hijiki, hiziki, ateji

Hijiki is an extremely popular seaweed in Japanese cuisine, but there is some hesitation about it internationally because studies have shown that this species concentrates arsenic. The US, UK, and Canada have all issued advisories against consuming it. However, you would need to eat more than 4.5 grams of hijiki per day, every day, to reach harmful arsenic levels, which is over five times the average person's daily consumption of the seaweed in Japan. Further, this seaweed species is codified in one of the oldest Chinese herbal medicine texts—the *Shennong Bencao Jing*. It was specifically used for treating tumorlike growths, edema, and urinary troubles.

Current research indicates that compounds in hijiki demonstrate antitumor, immune-regulating, antiviral, and antioxidant effects, but, like many other seaweeds, human clinical trials are still limited.

Hijiki is a delicious and fun seaweed to cook with and enjoy in moderation. It needs to be soaked longer than other seaweed to rehydrate (at least 30 minutes), and heating it at 90°F (32°C) for at least 5 minutes reduces any arsenic present by up to 80 percent. It's high in iron, magnesium, calcium, and dietary fiber. You can find it dried in Asian groceries or order it online.

Hijiki No Nimono

SERVES 2

This is a variation of a classic preparation of hijiki. It's often called a salad, but all the ingredients are simmered and then served at room temperature. I like to add scallions and ginger, and then mix it with quinoa for a full main dish.

INGREDIENTS

¼ cup dried hijiki

2 cups water, plus more for soaking

½ cup shelled edamame (frozen or fresh)

1 cup quinoa

1 tablespoon toasted sesame oil

1 large carrot

3 scallions, chopped

¼ teaspoon grated fresh ginger

⅔ cup dashi broth (see page 90)

2 tablespoons soy sauce

1 tablespoon rice wine vinegar

1 teaspoon maple syrup

1 teaspoon mirin

Optional toppings: cilantro, sesame seeds, squeeze of lime, chili crisp

INSTRUCTIONS

1. Soak hijiki in cold water for 2 to 4 hours (up to overnight). Drain and rinse twice.

2. Boil edamame until tender, 5 to 7 minutes. Drain and set aside.

3. Prepare quinoa: Bring 2 cups of water to a boil, add quinoa, and cook with the lid off until all the water has been absorbed. Fluff with a fork and set aside to cool.

4. Heat oil in a skillet over medium-high heat. Add carrot and cook until it starts to soften, 3 to 5 minutes.

5. Add hijiki, scallions, and ginger. Cook for 2 to 3 minutes.

6. Add dashi broth, soy sauce, vinegar, maple syrup, and mirin. Reduce the heat to medium and simmer until almost all the sauce is reduced, 7 to 10 minutes.

7. Remove the pan from the heat. Toss edamame with the carrot mixture, then combine everything with the prepared quinoa.

8. Enjoy warm, or chill for 2 hours up to overnight in the fridge and serve cold. Add toppings as you desire.

Arame that has been sliced, dried, and rehydrated for cooking. Arame is sweet with a gentle umami flavor.

Arame *Eisenia bicyclis* or *Ecklonia bicyclis*

Phaeophyceae
COMMON NAMES: sea oak, arame

Arame lives in temperate Pacific waters and is widely consumed in Japan. It's a mild and slightly sweet-tasting kelp. It's high in calcium, iron, magnesium, and vitamin A. There are notably high concentrations of lignan, alongside the typical gel-forming compounds (laminarans and alginates) in other brown seaweeds. Arame extracts have demonstrated anti-inflammatory, antioxidant, anticlotting, hypoglycemic, hypolipidemic (cholesterol-lowering), and neuroprotective actions and have most recently been examined as a topical inhibitor of MRSA infections and a photoprotective agent to reduce long-term skin damage from UV exposure. The following recipe highlights the texture and gentle umami flavor of this seaweed. You can usually find it dried in grocery stores from the Emerald Cove brand, or you can order it online.

Arame and Burdock Pasta

SERVES 2

I know seaweed on pasta might sound weird, but trust me on this one.
My partner's aunt used to make us a variation of burdock and seaweed
with pasta whenever we visited her, and those memories have inspired
this version.

INGREDIENTS

½ cup loosely packed dried arame

1 fresh burdock/gobo root, peeled and sliced
into thin matchsticks (about ¾ cup total)

½ pound linguine or fettuccine

1 tablespoon white miso paste

2 tablespoons olive or other neutral oil

3–5 scallions, diced

2–3 cloves garlic, pressed or finely chopped

1 teaspoon rice wine vinegar

Lemon, for serving

INSTRUCTIONS

1. Place arame in a bowl with cool water for
   30 to 60 minutes to rehydrate. Drain and rinse
   twice; set aside.

2. Immediately after peeling and slicing
   burdock root, place it in a bowl of cold water
   for 15 to 30 minutes. Soaking helps keep the
   root from oxidizing and turning brown, and it
   also removes some bitterness.

3. Cook the pasta according to the directions
   on the package. Drain, reserving ⅔ cup of the
   cooking liquid in a small bowl, and set pasta
   aside.

4. Add miso to the reserved pasta water and stir
   to dissolve.

5. Heat oil in a nonstick pan over medium heat
   and add burdock and scallions. Sauté until the
   root starts to soften, about 10 minutes.

6. Add garlic and sauté for an additional minute.

7. Stir in miso and pasta water along with vinegar, and simmer until the liquid has reduced
   some, 7 to 10 minutes.

8. Add arame; simmer for 2 minutes more.

9. Remove from the heat and mix in pasta. Serve
   immediately with a squeeze of fresh lemon.

This fresh green caulerpa is high in protein and is best eaten raw.

Caulerpa *Caulerpa lentillifera*

Chlorophyta
COMMON NAMES: sea grapes, green caviar, ume-budo (Japanese), lato (Korean)

Caulerpa is a large genus, with species that take many forms. *C. lentillifera* truly do look like miniature bunches of grapes that are the typical glowing green of Chlorophyta. They grow in tropical regions, in shallow waters with sandy or muddy bottoms. *C. lentillifera* is high in fiber, though less so than *Ulva* spp. It contains all the amino acids except tryptophan, and it has weight-to-volume protein levels on par with soy and eggs. (That said, seaweeds are so light that in order to get adequate protein, you would need to eat such a large volume that you would run the risk of overdosing on other vitamins or minerals.) There are appreciable levels of potassium, magnesium, and calcium as well as vitamins A, B2, and C, and polyunsaturated fatty acids, including alpha-linolenic acid. Like other green seaweeds, *Caulerpa* is low in iodine. Clinical research (mostly animal and in vitro studies) has demonstrated antioxidant, anti-inflammatory, cardioprotective, anticancer, hypoglycemic, hypolipidemic, and antimicrobial effects of *C. lentillifera* and its extracts. You can find sea grapes fresh in markets in regions where they are cultivated, but in international markets they are most often sold dried in brine. A tiny pinch will rehydrate at least five times its dried size. The "grapes" are like miniature salty tapioca pearls, adding a flavor pop to any dish you garnish them with (hence the common name *green caviar*). This is one seaweed that you actually do not want to cook—high temperatures destroy the texture that you are after. I recommend adding it to sushi or dipping the bunches in a sweet citrus and soy sauce for a full flavor experience.

Umibudo Sushi

MAKES 6 ROLLS

I would be remiss not to include sushi in a book about seaweed! Making your own sushi is really easy. I recommend investing in a bamboo mat to help with rolling. I like to make my sushi with seaweed on the outside (maki style), so that is what I offer you here. Take the extra time to season the sticky rice and chop your vegetables thinly so that they roll nicely. You can also lightly toast the nori sheets before rolling by laying them in a hot nonstick pan for 30 seconds on each side.

INGREDIENTS

UMIBUDO
1 teaspoon dried umibudo, or about
   2 tablespoons rehydrated
2 tablespoons soy sauce
1 teaspoon rice wine vinegar
1 teaspoon yuzu juice or 1 tablespoon lemon juice

RICE
1 cup sushi rice
3 tablespoons rice wine vinegar
1 tablespoon mirin

ROLLS
6 sheets nori
1 avocado, thinly sliced
Half a cucumber, thinly sliced into matchsticks
1 plum, thinly sliced

FOR SERVING
Soy sauce
Wasabi
Pickled ginger

INSTRUCTIONS

1. Prepare umibudo: Soak seaweed in a bowl of cool water for several hours (up to overnight) to fully rehydrate. Change the water at least once during the soak to help remove brine. Once rehydrated, rinse once more, and use scissors to snip into small groups of "grapes" that will top each piece of sushi.

2. Mix soy sauce, vinegar, and yuzu juice. Add umibudo and let the mixture marinate while preparing the rolls.

3. Make sushi rice according to the package instructions. Once done, remove from the heat and stir in vinegar and mirin. Let sit with the lid off to cool a bit.

4. Lay one sheet of nori on a bamboo mat.

5. Spread a thin but full layer of rice across nori, leaving 1 inch free of rice at the top edge.

6. Arrange your fillings in rows on top of the rice, covering just the bottom quarter of the nori.

7. Carefully start rolling the bottom edge over the fillings, tucking the edge under, and continuing to roll until the final edge. Dip your finger in water to gently wet that last seaweed edge and finish the roll.

8. Use the mat around the sushi roll to press firmly around the roll.

9. Remove the roll from the mat; let it sit for a couple of minutes, allowing nori to absorb some moisture from the rice. Cut the roll into rounds for serving. Top each piece of sushi with some umibudo.

10. Serve the sushi with soy sauce, wasabi, and pickled ginger. Rolls will keep in the fridge for 1 to 2 days.

EPILOGUE

You have ventured deep into the tide pools now, and as you return to dry land, I hope you have a frond of bladderwrack in your pocket, a leaf of sea lettuce on your tongue, and some laver in your bag. Each one has a unique flavor, texture, and place in your kitchen and medicine cabinet.

Seaweeds are organisms that are overlooked in the ecological web by so many of us; we often see them in their states of decay, or static in a photo, and we do not recognize how they have made their way into our food, beverages, and pharmaceuticals. Our human lives are intertwined with seaweeds, whether we look to them for medicine, for food, for fuel and building material, or simply for the oxygen we breathe. Seaweed farmers, harvesters, and wild-stock caretakers will need to be as much a part of our future food system as land-based agriculture. And we must remember the power of relationship; so much is possible when we see plants and seaweeds as allies for our physical, emotional, and spiritual well-being. By picking up this book, you have begun building a relationship with seaweed, and I hope I've inspired you to continue to honor and care for our necessary ties to the sea.

I would like to leave you with one last seaweed in these pages—a rather unusual red algae that I have known a long time and that holds a special place in my personal constellation of plant relationships.

Coralline algae "skeletons" made of calcium carbonate

Red maerl

The road that I lived on while growing up washed out more than once during hurricane season. Once the waves receded, trucks would dump loads of what my parents called marl into the holes. This was a magical time, because this road fill consisted of fossilized shells and coral chunks that provided hours of excavating activity. Some years, you could find huge intact cowrie shells and alphabet cones, or what we called yellow diamonds—honey calcite crystals that form inside fossilized clam shells. I later learned that the term *marl* actually refers to a sediment layer (and crucial habitat) mostly found in Florida's Everglades region. This layer, formed as dissolved calcite (from calcium carbonate), precipitated as crystalline "needles" in mats of cyanobacteria, green algae, and diatoms. This word *marl* also exists across the Atlantic, where the term *maerl* refers to large beds of specific coralline algae.

Coralline algae are architects, slowly but surely building their bodies and constructing their world. They create vast structures and communities out of our sight, always submerged in the sublittoral zone. They quite literally are the living cement that is the base of coral reef ecosystems, and they hold together the spaces in between corals. They are also some of the most vulnerable species—like corals, their calcium carbonate skeletons start dissolving as the surrounding water acidifies. We don't know what the future of coralline seaweeds is, but my guess is that they will disappear in some places, reappear in others, and continue adapting because algae are some of the most resilient organisms on this planet. Seaweeds can migrate as much as land plants, animals, or humans.

Coralline algae ask us questions about what we are building. What relationships can we tend with the sea, the soil, and the plants? What are the strengths that we create on a cellular level? What memories live in our bones; those that were given to us, and those that we pass on and leave behind?

I know the memory of salt water lives in my blood, created from my bones; and I look to seaweeds for advice on how to be a good ancestor. I see the ways they celebrate queerness with shapeshifting morphology and wildly creative sex lives. They teach embodiment that is solid and strong and rooted, but they also have constant flex and response to change. I see how seaweeds organize in the intertidal, each species occupying their ideal niche to maintain the health of the ecosystem. They can be competitive and opportunistic, but inevitably they create habitat and resources for other organisms. And in their decay, they return their hyperconcentrated nutrients to the earth and the water, releasing and redistributing all that belonged to them during their lives, leaving memories of abundance and nourishing what is to come. Seaweeds contain the salt and minerals of our blood, the calcium of our bones, the slip and slime of our mucous membranes and soft tissues, and the watery flow of our lymph. We are them and they are us. Let them welcome you home.

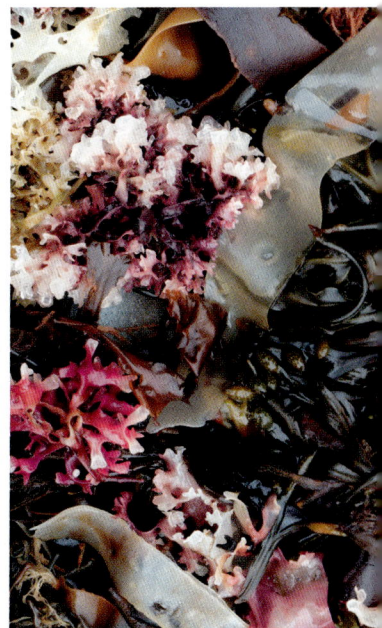

Irish moss, bladderwrack, and kelp tangled on the shore

MATERIALS

- Seaweed of choice
- Glass jar
- Vinegar of choice
- Parchment paper
- Cheesecloth, muslin, or other straining cloth
- Raw honey (optional, for oxymel)

METHOD

1. Cut, crush, or powder seaweed as small as possible to maximize surface area for extraction. Add it to a jar, leaving 1 to 2 inches of space below the rim to allow seaweed to expand as it absorbs liquid.

2. Add enough vinegar to cover seaweed, leaving a small amount of room at the top of the jar. Place a piece of parchment paper on the jar mouth, then close the lid tightly and shake to make sure all pieces of seaweed are saturated.

3. Let the jar sit for 2 to 4 weeks, shaking as often as you remember.

4. After vinegar has steeped, strain it through cheesecloth, squeezing as much liquid out of the plant material as you can.

5. Decant vinegar into a clean jar or bottle and label it. If making an oxymel, add 30 to 100 percent as much honey as you have vinegar (e.g., if you have 8 ounces of vinegar, add 3 to 8 ounces of honey) and stir or shake to mix well.

6. Store the infused vinegar in the fridge for the longest shelf life, up to 6 months. Raw vinegar will continue to grow the mother at the bottom of the jar, but growth of this goopy, stringy material is totally normal.

Syrups

A syrup is a medicinal preparation that combines a decoction (long-simmered herbal infusion) with some kind of sugar to preserve it. This can be cane sugar, coconut sugar, sucanat, or raw honey. For the longest shelf life, use an equal volume of sweetener to decoction. If you are keeping it in the fridge or using it short-term for an acute condition, use less sweetener. Add a splash of vinegar or brandy to increase shelf life. Syrups are fantastic for kids and elders because they make bitter herbs taste good! This preparation is particularly suited for fruits and fresh plants that have a high water content. Seaweed is lovely in syrups because the hot water–soluble gel compounds are extracted and create silky textures. Unlike with tinctures and alcohol extracts, I recommend adding other herbs or fruits to a seaweed syrup. You can really play with flavors, and seaweed can be just one textural and taste component.

MATERIALS

- Seaweed of choice
- Roots and/or berries
- Leafy and/or aromatic herbs, spices, or flowers
- Medium-sized saucepan
- Water
- Cheesecloth, muslin bag, or other straining cloth
- Sweetener of choice

METHOD

1. First prepare the decoction component. Place seaweed, roots, and berries (if using) in a medium-sized saucepan with water in the amount that is 6 times the volume of the herbs in the pan. The large volume of water is because seaweed will expand 3 to 5 times its size once rehydrated. Roots and berries absorb large amounts of liquid as well, and

you want to keep everything submerged in the decoction even as the liquid is absorbed.

2. Bring the seaweed mixture to a simmer and cook, uncovered, for 30 to 40 minutes.

3. Add the more delicate herbs and spices, such as aromatic leaves or flowers, if using. Continue to simmer for 20 minutes. This short simmer allows them to be decocted while keeping flavors and aromas intact.

4. After 50 to 60 minutes of simmering, the water volume should be reduced by half. Remove the pan from the heat and let the mixture cool slightly. Strain it through cheesecloth, squeezing as much liquid out of the plant material as you can.

5. If using granular sugar, measure out an equal weight to the volume of decoction liquid (e.g., for 100 mL of liquid, use 100 g of sugar), add to the decoction, and stir to fully dissolve. If using honey, measure out an equal volume to the volume of decoction liquid (e.g., 100 mL of honey for 100 mL of decoction), add to the decoction, and stir to fully dissolve. When using raw honey, make sure the liquid portion is at or below 100°F (38°C) so beneficial enzymes in the honey are not destroyed.

6. Decant the syrup into a clean jar and label it. Store it in a cool, dark place for up to 3 months or in the fridge for up to 6 months.

Infused Oils

Infusing herbs into oil is a fantastic way to bring their medicine to your largest organ—your skin! Infused oils are the base ingredient for homemade seaweed skin creams (see page 216), and you can create these base oils with all sorts of other lovely aromatic and medicinal herbs. Some classic skin

herbs include calendula, plantain, yarrow, St. John's wort, rose, and lavender. You can use a variety of oils. I recommend sunflower or light olive oil as a relatively neutral base, but avocado, coconut, jojoba, and almond oils are lovely, too. Herbs can be infused fresh or dried but dried is better, as this will dramatically reduce the risk of mold forming (from water content) in the oil. Heating the oil is very important for extracting medicinal compounds and aromatics. An infused oil can be made in a slow cooker on low for 12 to 36 hours. You can set the jar on a wood-burning stove in winter or place it in a paper bag (to prevent UV exposure) out in the sun in summer for a few days. As with infused vinegars, I almost always make infused oils using the folk method.

MATERIALS

- Dried plant material of your choice
- Widemouth jar
- Oil of your choice
- Cheesecloth, muslin bag, or other straining cloth

METHOD

1. Place herbs in a jar, no more than three-quarters full.

2. Fill the jar with oil so that all of the herbs are covered. Most leafy herbs and flowers will float to the top. This is okay. Shaking or stirring will help disperse them.

3. If using the concentrated heat slow cooker method, fill the cooker at least halfway with water, and place the jar in the water. Set the cooker to the lowest temperature and let the jar sit in the hot-water bath for 12 hours to 4 days. (If steeping for multiple days, turn off the cooker periodically for a break, and remember to add more water as it evaporates.)

4. If using the solar method (or another more diffuse heat source), let oil infuse for at least 1 week, and up to 1 month.

5. Strain oil through cheesecloth, decant it into a clean jar, and label it. Store it in a cool, dark place for up to 1 year.

Salves

A salve is simply an oil combined with beeswax (or candelilla wax if you are not using bee products) to achieve a semisoft texture and increased shelf life. I've given you a basic recipe, but you can experiment with adding in other oils (e.g., rosehip, jojoba) or fats such as shea butter or tallow. A few drops of essential oil will prolong the shelf life. My preferred salve texture is a ratio of 1:8 beeswax to oil, but you may have a different preference for hardness.

MATERIALS

- 30 mg beeswax (or candelilla wax), in small pieces
- Double boiler or slow cooker
- Stirring utensil
- 240 mL infused oil of your choice
- Containers for finished salve

METHOD

1. Melt beeswax in a double boiler or cooker. Adding a small bit of oil to the beeswax helps the melting process go faster, as does frequent stirring.

2. Add all of the remaining oil to melted beeswax and continue stirring until fully mixed together.

3. Pour the hot salve into containers. Let cool with the lids off.

4. Label and store in a cool, dark place for up to 1 year.

Creams

One of the most decadent topical preparations is a cream. These are a little trickier than salves because a water component is part of the recipe. You can use a strong tea (well filtered), a carrageenan infusion, a hydrosol, or plain distilled water. A countertop blender makes the process easiest, but you can make a cream with an immersion blender or even a mortar and pestle and a lot of elbow grease. This recipe was passed to me by herb school teachers, originally adapted from Rosemary Gladstar's cream recipe. This makes 450 mL, or about 15 ounces, which tends to be the right size for the average blender. Play around to see what works for you!

MATERIALS

- Double boiler
- Stirring utensil
- Blender
- Containers

Oil Phase (53% of total volume)

- 240 mL oil (infused or plain, but must be a variety that is liquid at room temperature)
- Optional: 10–15 drops essential oil of your choice, 30 drops vitamin E oil

Water Phase (40% of total volume)

- 180 mL liquid (distilled water, hydrosol, infusion, aloe vera juice)

Hardener (7% of total volume)

- 30 g beeswax
- Optional: 5–10 grams cocoa butter, shea butter, or other solid fat/oil. If using, reduce the amount of beeswax by the amount of butter used, and know that the cream will be slightly less stiff when it cools.

INSTRUCTIONS

Note: Sterilize/sanitize your containers first! Boil or wipe the container and lid with 70% isopropyl alcohol.

1. Melt beeswax and other hard butters in the oil phase over low heat or in a double boiler.

2. Remove from the heat and let cool slightly, but don't let beeswax harden.

3. Add the water phase to the blender, turn it on high, and let it warm a bit (1 to 2 minutes). For best results, the temperatures of the oil-wax combination and the liquid should be close to each other.

4. If desired, this is the time to add essential oils or antioxidants (e.g., vitamin E oil).

5. Blend on low while slowly adding the oil-wax combination in a very thin stream to the liquid; increase blender speed as you add more.

6. When the blender "chokes" or catches, stop it and scrape down the edges with a spatula. Turn the blender back on, and continue blending and scraping the sides until the cream is fully whipped and emulsified.

7. While still warm, pour the cream into containers.

Lotions

For lotions, the water phase is larger than the oil phase, which makes emulsification trickier. You'll need to melt and add some emulsifying wax (this can be bought online or in some natural products stores) instead of another hardener to keep everything creamy.

RECIPE RATIOS FOR LOTIONS

- Water phase: 73–75% total volume
- Oil phase: 25–27% total volume
- 1% emulsifying wax (about 5 g per 100 mL lotion)

Flower and Plant Essences

One of my favorite ways to partner with plants is by using essences. Bach's flower essences are probably the most famous version of this medicine. Essences are energetic medicine, vibrational imprints of a plant or place that are preserved in water and either alcohol, vinegar, or glycerine. I love essences because they are both personal and universal. The medicine of a plant carries a common story, even as each person working with it will create a unique essence. They are also very safe; since there is no physical plant in the medicine, they can be taken alongside any medications and are safe for children, elders, and pets. While these were originally called flower essences, they can be made with seaweeds and nonflowering plants, and I have made incredibly powerful and deeply nourishing essences with them.

Here is a basic method for making an essence, but please use your intuition! The plants will tell you how to partner with them and how to make the right medicine for the moment.

MATERIALS

- Vessel for steeping essence. Traditionally, clear glass is used, but you can use any small vessel that can hold water and that's meaningful to you and your practice.
- Distilled water
- Strainer (optional)
- Jar/container for finished essence
- Preservative of your choice. Brandy is traditional, but you can use any alcohol, vinegar, or vegetable glycerine.

METHOD

1. Choose the plant you'd like to make an essence of—or rather, let it choose you. Just like harvesting, ask permission of the plant to make medicine with its energy. If you receive a yes, offer a thank-you in exchange.

2. Fill your essence vessel with distilled water.

3. Place plant material in water. For plants with flowers and leaves, you might choose to harvest a few, or use the no-pick method and place the vessel so that the plant hangs above or in the water. You can also simply place the vessel beside or underneath the plant. For seaweeds, you may choose to harvest the plant, or simply place the vessel, filled with either distilled or ocean salt water, in a tide pool or along the shore, such that the seaweed is in the water.

4. Let sit for at least an hour, and up to as long as the plants and your intuition tell you to! Edward Bach said you must make essences in full sunlight, but I have found that clouds, changing weather, and moonlight all add potent medicine to an essence as much as sunlight does. For seaweeds, I like to let the essence sit as the tide goes out and then comes back to the place where the essence is steeping.

5. Once the essence has steeped, remove or strain out plant material.

6. Decant the water into the container for your "mother" or "parent" essence, filling halfway. Fill the remainder of the jar with the preservative of your choice. Label and store in a cool, dark place. For further dilution and dosing, see the next section.

DILUTION AND DOSING

Dosing flower and plant essences is different from other physical extracts. The principle is similar to homeopathy, which holds that the lower the dose of medicine, the greater the effectiveness. We generally use drop doses (rather than dropperfuls or teaspoons), as a little bit goes a very long way. Energetic medicine is subtle medicine, and so we dilute our original essence for dosing. The original essence that you make and preserve is often referred to as the "mother" essence, though I prefer "parent" or "primary," as it's less gendered. From the primary essence, you dilute further into a "stock bottle," and further still into a "dosage bottle."

To make a stock bottle, fill a bottle halfway with a solution of 50 percent water, 50 percent preservative. Then fill the remaining half of the bottle with some of the parent essence. Finally, to make a dosage bottle, fill your bottle with a solution of 50 percent water and 50 percent preservative. To this, add a number of drops of your stock essence. The number is up to you and your own intuition. Now you have a suitably diluted flower or plant essence!

HOW TO USE AN ESSENCE

Flower and plant essences can be consumed directly from the dropper or added to a beverage. You can add some drops to a bath, or place a few on your wrists, temples, or any other body part that feels right. You can add some to a mister and spritz it around your space.

As for how many drops to use, this is again an opportunity to let your intuition guide you. A pretty standard range is 1 to 10 drops, but you can do what feels right. Play with it, listen, and the medicine will be there for you.

APPENDIX 2   Natural Dyeing with Seaweed

Alongside my herbal practice of interacting with plants as medicine, I also grow and harvest plants for natural dye. Natural dyeing has been experiencing a resurgence in recent years, with a growth in seed companies focused on dye plants, botanical pigment collectives, and large-scale community dye vat projects, which offer mail-in services to dye your clothes and ship them back to your closet.

The development and application of botanical pigments throughout history is rich and layered, but historical mention of pigments from the sea is scant. The most notable example is perhaps Tyrian purple. As early as the fifteenth century BCE, Phoenicians and others living on the coasts of the Mediterranean harvested hundreds of thousands of spiny dye murex (*Bolinus brandaris*). These marine snails have glands that secrete a mucus which, when exposed to air, turns from milky white to yellow to green and finally to a violet purple that forms a lasting dye on fabric. Today, researchers are looking at how marine animal and plant pigment compounds might provide replacements for synthetic colors, but that sort of high-tech biochemistry isn't exactly accessible in your home kitchen. Many folks are now experimenting with pigments derived from sea plants, and I have started my own color studies from seaweeds.

Natural dyeing is both a complex science and an experience of letting go of expectations. I always come into the practice with curiosity as my motivation: What will this plant harvested at this time of the year look like on textile? On paper? Transformed into a lake pigment? Seaweed tones are subtle, and I recommend experimenting with silk textile or quality watercolor paper. You can expect a range of pinks and browns, with some hints of gray and yellow. Here I will provide you with basic instructions for preparing fabric and creating dye with seaweed. Know that the process will be rather smelly, and your fabric will require multiple washes to not smell like low tide—but the process is part of the joy!

Murex shell

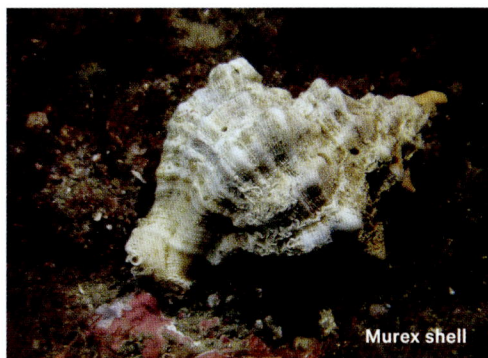

Silks dyed with bladderwrack and rockweed

MATERIALS

- Silk or wool fabric; protein-based fibers take up natural dyes without a two-step mordanting (pretreatment) process that would be required to work with cellulose-based fibers like cotton or linen
- Water
- Alum (aluminum potassium sulfate), sold at art supply stores or online
- Cream of tartar (a small amount added to the mordant vat keeps silk and wool fiber soft)
- Large nonreactive pot, such as stainless steel or enamel coated (not one you use for cooking!)
- Wooden or nonreactive metal spoon, or silicone spatula
- Several handfuls of seaweed of your choice (enough to fill your pot halfway), harvested responsibly (see Harvesting Techniques & Ethical Wildcrafting on page 31)
- Hot plate, propane stove, or other outdoor-use heating element (unless you don't mind your kitchen smelling intensely of low tide)
- Strainer/colander
- Bowl
- Mild soap

Silks being mordanted with alum

INSTRUCTIONS

Step 1: Preparing Your Fabric/Mordanting

Mordanting refers to the process of treating fabric so that it will take up dyes. Typically this involves the use of metal salts (such as alum) to chemically bind pigment to fabric.

1. Weigh your fiber/silk while it is dry. Then place in room-temperature water to soak.

2. Measure alum at 12% of your weight of fiber (WOF). Weigh out an equal amount of cream of tartar.

3. Place alum and cream of tartar in the nonreactive pot and add hot water to almost fill the pot. Stir to dissolve alum and cream of tartar completely.

4. Add the fiber to the alum vat and stir so that all parts of the fiber are in contact with the solution.

5. Let sit for 4 to 12 hours.

6. After soaking the fiber, remove it from the alum vat and hang to dry. *Do not rinse until completely dry*. This allows the alum to fully bind to the fibers. The alum solution may be disposed of down the kitchen sink.

Dye pot filled with bladderwrack and a silk in process

The tones resulting from dyeing with seaweeds are subtle, often pink and brown with hints of gray and yellow.

Step 2: Creating a Dye Vat

1. Place seaweed in the same nonreactive pot and cover it with water.

2. Heat seaweed vat to a gentle simmer (try not to boil it!). Simmer with the lid on for at least an hour. (Yes, it will smell!) Turn off the heat and let sit for approximately 12 hours, though you can experiment with the steeping time—this is just a guideline for letting color develop.

3. Strain seaweed out of the vat, if desired. Your pot of dye is now ready for fabric! (You can also keep the seaweed in the dye vat as shown above; however, the color may be uneven, as plant material presses on parts of the fabric.)

Step 3: Dyeing Your Fabric

1. Soak dried mordanted fiber in a bowl of water to fully moisten. This presoak makes it easier for color to permeate the fabric, and it reduces the chance of temperature shock when adding a natural fiber to hot water.

2. Add the fiber to the vat with the strained water and heat again, keeping the temperature below 100°F (38°) to preserve the texture of silk or wool. Keep the vat warm (either keep on low or turn off the heat) for at least an hour. You can also let the fiber sit in the pot off the heat for another 12 hours.

3. Remove the fiber from the vat and let it dry fully before washing in cold water with mild soap.

Seaweeds & Where to Source Them

Seaweed is becoming more and more available commercially, which is amazing! Seaweeds are often available dry, and you can find frozen puréed kelp cubes in Whole Foods Market or other natural grocery stores. Seagrove Kelp (based in Alaska) sells flash-frozen kelp in bulk in various grocery stores on the West Coast of the US. You will usually find dried seaweeds in the "Asian" or "International" section of the grocery store, even though many of them are grown and harvested in the US. If you have an Asian grocery in your town, you will be able to find some other types as well. Agar is sold in flake form or as a powder. Grocery store bulk sections sometimes carry it, or you can find it next to the dried seaweeds. Nori sheets are made by many companies and are widely available. The kelp noodles from Sea Tangle Noodle Company are fun to experiment with!

Following is a list of seaweed farms and wild harvesters in the US and the seaweeds they usually have available.

Seaweed Farms and Wild Harvesters

Atlantic Holdfast (ME): atlanticholdfast.com
Alaria, *bladderwrack, dulse, Irish moss, Atlantic kombu (Laminaria digitata), sugar kelp*

Ironbound Island (ME): ironboundisland.com
Alaria, *bladderwrack, dulse, laver, Atlantic kombu (Laminaria digitata), sugar kelp*

Maine Coast Sea Vegetables (ME): seaveg.com
Alaria, *bladderwrack, dulse, Irish moss, laver, rockweed, sea lettuce, sugar kelp*

Maine Seaweed (ME): theseaweedman.com
Alaria, *dulse, Atlantic kombu (Laminaria digitata), sugar kelp*

Nautical Farms (ME): nauticalfarms.com
Sugar kelp

Rhody Wild Sea Gardens (RI): rhodywildseagardens.com
Sugar kelp (fresh and dried)

Seagrove Kelp (AK): seagrovekelp.com
Ribbon kelp (Alaria), sugar kelp (fresh and dried)

Stonington Kelp Co. (CT): stoningtonkelpco.com
Sugar kelp

Seaweed Products

Atlantic Sea Farms (ME): atlanticseafarms.com
Dried kelp, fresh frozen kelp, seaweed-based condiments, seaweed veggie burgers

Barnacle Foods (AK): barnaclefoods.com
Condiments from hot sauce to jam, skincare products

Cup of Sea: cupofsea.me
Delicious seaweed teas

Daybreak Seaweed Co. (CA): daybreakseaweed.com
Seasoning blends, condiments

Dulse & Rugosa (ME): dulseandrugosa.com
Seaweed-based skincare

Heritage Seaweed (ME): heritageseaweed.com
Brick-and-mortar location run by the founders of Cup of Sea, stocking bulk dried seaweeds as well as seaweed products and literature from around the country

Springtide Seaweed (ME): springtideseaweed.com
Spice blends, soup blends

Herbs

The recipes in this book include a variety of medicinal herbs. If your local grocery store has a bulk herbs and spices section, you can find many of them there. If not, you can order herbs online. I feel very strongly about sourcing herbs grown regionally, or from farms where there is transparency about the growing and harvesting processes *and* labor practices. Here are my recommendations:

Cutting Root Farm (PA):
cuttingroot.com

Foster Farm Botanicals (VT):
fosterfarmbotanicals.com

Foxtrot Herb Farm (MA): foxtrotherbfarm.com

Free Verse Farm (VT): freeversefarm.com

Golden Hour Farm (MI): goldenhourherbs.com

Meeting House Farm (ME): meetinghouse.farm

Oshala Farm (OR): oshalafarm.com

Wayside Botanicals (WA): waysidebotanicals.com

For Chinese herbs, you will likely need to order from a larger company that sources herbs internationally. I recommend Pacific Botanicals.

Pantry Staples for Cooking and Making Medicine with Seaweed

Keep these ingredients on hand for a well-stocked pantry. You can find most of these at your local grocery store or Asian market.

Beans

In my opinion, beans should always be cooked with a little seaweed. And if you are going to go to the effort of cooking and deliciously spicing dry beans, you should use excellent beans! Rancho Gordo is my favorite bean company, based in Napa, California. They grow dozens of heirloom varieties and partner with farms in Mexico to cultivate and preserve rare bean varieties.

Gochugaru and Gochujang

Gochugaru are Korean chile flakes, and gochujang is a fermented paste of these chiles with rice starch and other ingredients. The flakes themselves are pretty mild and have a slightly fruity taste. The paste is much more concentrated spice and adds a distinctive kick to dishes.

Mushrooms

Seaweed and fungi go hand in hand as best friends in the world of umami. Find fresh mushrooms at your local grocery. Asian groceries usually sell large bags of dried shiitake mushrooms, which are perfect for making batches of seaweed and mushroom dashi. You can also dry your own, gill side up in sunlight to maximize their bioavailable vitamin D.

Oats

While we often associate seaweed with the flavors of Asian cuisine, when we start cooking traditional European recipes, other ingredients come forward. Rolled oats are a primary one. The recipes in this book use rolled oats, not quick-cook or instant. You can usually buy these certified gluten-free as well.

Oils

Sesame oil is the most traditional choice to pair with seaweed in Asian dishes. You can quickly sauté seaweed in it before adding to soup, or use it in dressing for seaweed salad. It is sold plain or toasted, and quality varies quite a bit; taste different brands and see what you like! For recipes that call for another neutral oil, use olive oil.

Sesame Seeds

Sesame seeds are as important to pair with seaweed as the oil. They can be black or white, and toasting brings out the sweet, nutty flavor. Tahini (ground sesame seeds) makes a great base for dressings for savory seaweed dishes, or it can be incorporated into baked goods.

Soy Products

Seaweed and soy have been matched in Japanese culinary traditions for centuries. Let's start with whole soybeans—or edamame. Fresh or frozen, they are one of my favorite vegetarian dumpling fillings, and they mix well into salad or stir-fries. Fermented soybeans in the form of tempeh are a fantastic base for seaweed spice mixes and marinades, and tofu swims alongside seaweed in soups. The mild flavor of both allows them to bring substance and protein to a dish while letting the seaweed shine. Finally, soy sauce is a sauce, a marinade, and a topping for seaweed. This fermented condiment is quite a bit more varied than the ubiquitous Kikkoman bottles. Dark and light soy sauce have different applications and flavor profiles, and you can use regular or reduced salt in any of the recipes in this book. If you are sensitive to gluten, substitute with tamari (soy sauce made with rice rather than wheat).

Spices

Your food is only as good as your spices! A strong statement, but quality spices truly elevate a dish. Several recipes call for guntar sannam chiles, which are best sourced from Diaspora Co., an incredible queer- and woman-owned spice company now based in India and dedicated to single-origin, same-year spices. I especially recommend them for chiles, cinnamon, black pepper, coriander, fennel, cumin, cardamom pods, and turmeric.

Tamarind Paste

The sour tamarind fruit comes from Africa and has naturalized in Asia. Its sweet and tangy pulp is blended into a purée that adds a burst of flavor to sauces and marinades—and complements the salty umami of seaweed so well. It's sort of a cross between balsamic vinegar, lemon, and dates.

Vinegars

When cooking with deeply umami seaweed, salt and sour are natural taste pairings. Soy sauce often fills the salty note, and the sour usually comes from vinegars. Rice wine vinegar is generally my first choice—it's mild and a little sweet, but it also has this mineral-y quality that goes so well with seaweed. Apple cider vinegar is also key—especially for many of the herbal recipes that involve infusing herbs and seaweed into raw vinegar. Bragg is the brand sold everywhere, but you can also make your own by fermenting apples. Apple cider vinegar, and any fruit vinegar, will have flavor notes corresponding to the fruit from whence it came, which is an exciting way to add subtle taste to a dish or herbal medicine.

Glossary

Adaptogen/adaptogenic: An herb/herbal action that helps increase a body's ability to adapt to and recover from stress; generally has effects on multiple body systems with particular affinity for one or two

Alginate: A polysaccharide found in all brown algae

Alterative: An herb/herbal action that has a normalizing or balancing effect on all body systems, specifically by strengthening the pathways of assimilation and elimination (e.g., liver, kidneys, lungs, and lymph)

Angiotensin-converting enzyme (ACE): An enzyme that is part of the renin-angiotensin system that helps regulate blood pressure by causing blood vessels to constrict and increase pressure

Anthelmintic: Antiworm (herbal action)

Antioxidant: A substance that inhibits oxidation in the body and reduces oxidative stress, which refers to cell and DNA damage caused by reactive oxygen species (normally occurring, but highly reactive, molecules from oxygen metabolism) in the body

Ascophyllan: A polysaccharide present in *Ascophyllum nodosum*

Carminative: Gas relieving, bloat reducing (herbal action)

Cytokine: A type of small protein secreted by immune cells to send signals to other cells in the body

Demulcent: Soothing to internal tissues, usually applicable to mucilaginous plants (herbal action)

Dichotomously: Dividing into two parts

Diploid: A cell containing two distinct sets of chromosomes, one contributed from each parent

Discoid (holdfast): Disc-shaped holdfast

Edema: Swelling caused by fluid buildup in tissues

Emollient: Soothing to the skin (topical herbal action)

Epiphyte: Growing on another plant but not taking the host plant's resources

Frond: The equivalent of a leaf in seaweeds, synonymous with *blade*

Fucoid: Relating to, or resembling, brown seaweed

Fucoidan: A sulfated polysaccharide present in brown seaweeds

Gametophyte: The life-cycle stage of a seaweed wherein the reproductive structures produce gametes (sperm and egg cells)

Haploid: A cell containing a single set of chromosomes (e.g., a sperm or an egg cell)

Hepatoprotective: Protective of the liver cells (herbal action)

High-density lipoprotein (HDL): A molecule that absorbs cholesterol in the blood and carries it back to the liver to be flushed from the body; high levels of HDL are desirable

Holdfast: The specialized structure of seaweed cells that holds the algae in place on a rock or in substrate; a rough equivalent to roots in land plants

Hydrocolloid: A substance that forms a gel in the presence of, or when mixed with, water; seaweed-specific hydrocolloids are called phycocolloids

In vitro: A study or process conducted in a test tube, petri dish, or other setting outside of a living body

In vivo: A study or research process conducted in the body of a living organism

Laminaran: A carbohydrate storage molecule in brown algae

Littoral: (*n*) The region next to the shore, or (*adj*) related to the shore; often used interchangeably with *intertidal*

Low-density lipoprotein (LDL): A molecule that absorbs and circulates cholesterol around the body but does not remove it; low levels of LDL are desirable

Lymphatic: Promotes secretion and flow of lymph fluids (herbal action)

Mariculture: Ocean- or saltwater-based farming, in contrast to aquaculture, which is fresh water based

Metabolic syndrome: A diagnostic term used to describe a constellation of symptoms including high blood pressure, dysregulated blood sugar and cholesterol, and weight changes that increase risk of heart attack, stroke, and developing type 2 diabetes

Natural killer cell: A type of white blood cell responsible for killing tumor cells or virus-infected cells

Nervine: Calming or relaxing to the nervous system (herbal action)

Pelagic: Free-floating in the upper water column

Phycobiliprotein: Water-soluble proteins found in marine algae and cyanobacteria; *phycobilin* refers to any of the red or blue photosynthetic pigments in algae

Phycology: The study of seaweed

Pneumatocyst: Air-filled vesicles present on or as part of a seaweed thallus, serving as a buoyancy aid

Polysaccharide: A carbohydrate whose molecules consist of a number of sugar molecules bonded together; in seaweeds, this includes ulvan, carrageenan, fucoidan, porphyran, and ascophyllan (A sulfated polysaccharide means that there are sulfate groups within the molecular structure, which give these compounds unique biochemical properties.)

Porphyran: A polysaccharide that is characteristic of the *Porphyra* genus

Sori: Reproductive structure(s) of a seaweed

Sporophyte: The life-cycle stage of a seaweed wherein the reproductive structures produce spores

Stipe: The equivalent of a stem in seaweeds, connecting the holdfast to the blades or fronds

T cell: A type of white blood cell (lymphocyte) that is part of the adaptive immune response

Thalassotherapy: The use of seawater (and seaweed) in cosmetic and health treatments

Thallus: The entirety of a seaweed body

Ulvan: A primary polysaccharide in Chlorophyta seaweeds

Vermifuge: Antiworm (herbal action)

Vulnerary: Wound healing (herbal action)

Wrack: A catch-all term for washed-ashore seaweeds, sea grass, and driftwood; a term frequently used in the common names of brown seaweeds, whether alive or washed ashore

Zygote: A diploid cell resulting from the fusion of two haploid cells

Metric Conversion Charts

WEIGHT

TO CONVERT	TO	MULTIPLY
ounces	grams	ounces by 28.35
pounds	grams	pounds by 453.5
pounds	kilograms	pounds by 0.45

US	METRIC
⅛ ounce	3.5 grams
¼ ounce	7 grams
⅓ ounce	9.45 grams
½ ounce	14 grams
1 ounce	28 grams
10 ounces	280 grams
16 ounces (1 pound)	454 grams
20 ounces	567 grams

VOLUME

TO CONVERT	TO	MULTIPLY
teaspoons	milliliters	teaspoons by 4.93
tablespoons	milliliters	tablespoons by 14.79
fluid ounces	milliliters	fluid ounces by 29.57
cups	milliliters	cups by 236.59
cups	liters	cups by 0.24
pints	milliliters	pints by 473.18
pints	liters	pints by 0.473
quarts	milliliters	quarts by 946.36
quarts	liters	quarts by 0.946

VOLUME, *continued*

US	METRIC
1 teaspoon	5 milliliters
1 tablespoon	15 milliliters
¼ cup	60 milliliters
½ cup	120 milliliters
1 cup	240 milliliters
4 cups (1 quart)	0.95 liter
4 quarts (1 gallon)	3.8 liters

LENGTH

TO CONVERT	TO	MULTIPLY
inches	millimeters	inches by 25.4
inches	centimeters	inches by 2.54
inches	meters	inches by 0.0254
feet	meters	feet by 0.3048

US (INCHES)	METRIC (CENTIMETERS)
⅛ inch	3.2 mm
¼ inch	6.35 mm
½ inch	1.27 cm
¾ inch	1.91 cm
1	2.54
1.5	3.81
2	5.08
5	12.70
10	25.40
15	38.10

Interior Photography Credits

Index

Page numbers in *italics* indicate images.

P

R

S